Seeing Trees:

A Poetic Arboretum

Seeing Trees: A Poetic Arboretum

BY

JOHN C. RYAN
& GLEN PHILLIPS

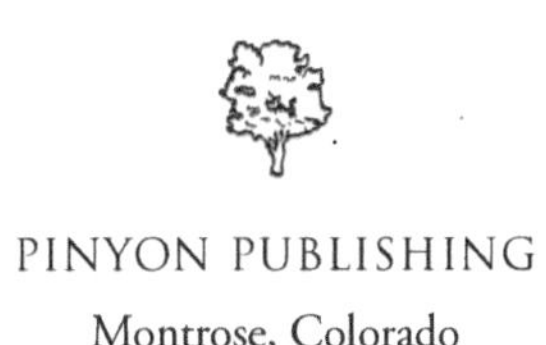

PINYON PUBLISHING
Montrose, Colorado

Cover and Interior Art by Glen Phillips

Photograph of John C. Ryan by Himself

Photograph of Glen Phillips by Lily Liu

First Edition: August 2020

Pinyon Publishing
23847 V66 Trail, Montrose, CO 81403
www.pinyon-publishing.com

Library of Congress Control Number: 2020940411
ISBN: 978-1-936671-67-0

ACKNOWLEDGMENTS

In 2018 the following poems were published in *In the Hollow of the Land: Collected Poems 1968-2018* (Wild Weeds Press, Perth WA):

Mrs. Dance's Jarrah Tree
Footsteps Among Trees
Wind Seed Song
The Balga
Unidentified Flying Image of a Poem
Dryandra Dreaming
Back to Sweeping My Paths
Charred Ground
When the Bush Burns
Acknowledgment of Jam Tree Country
Small Beauty and Granite Kunzea

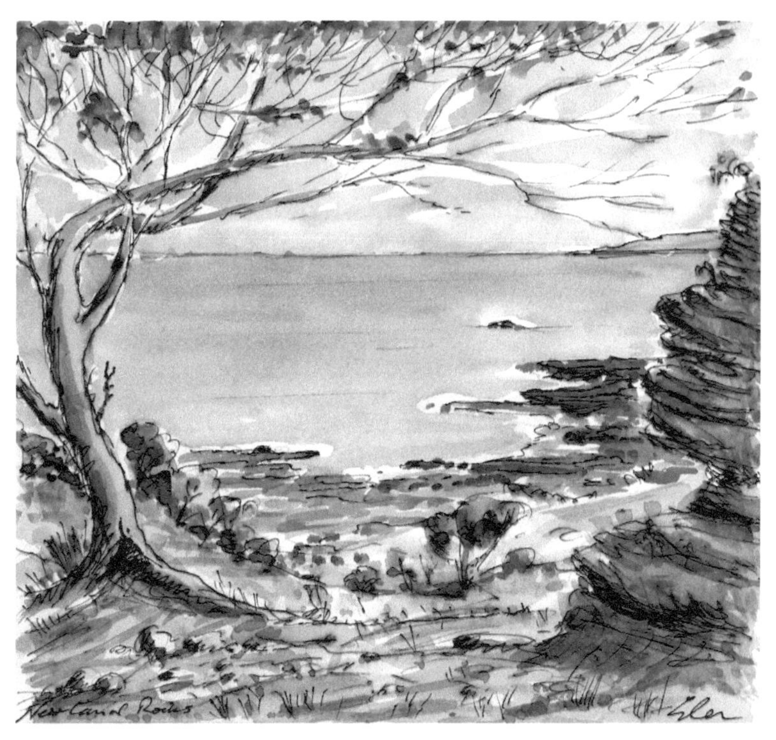

Powderbark Wandoo and Canal Rocks

Contents

I. Tree Signs

John C. Ryan

II. Tree Tales

Glen Phillips

III. Tree Forms

John C. Ryan

IV. Trees Burn

Glen Phillips

I. Tree Signs

John C. Ryan

INSIDE A JARRAH TREE, A BLACK TUNNEL REACHING SKYWARD

Jarrah (*Eucalyptus marginata*)

neatly burned-out innards,
this tree lives on as skin,
still supple and twisting in pleats,
but where did the heart go, and the breast bone
and the heavy, unctuous insides?

the spine endures, knobby column
ripped bare by a magnificent thrust of liquid fire—
but what about the soul,
where is its perch now?

outside, the grass trees don
verdant headdresses over the charred land,
and kino sap stamps red
insignias along marri gum trunks.

have you ever breathed inside a tree
to feel the cool glance of air
where once a molten river ran,
seeing the outside from within?

witchetty grubs or magpies might,
clawing skyward towards a portal of light

but I would not stand here forever.

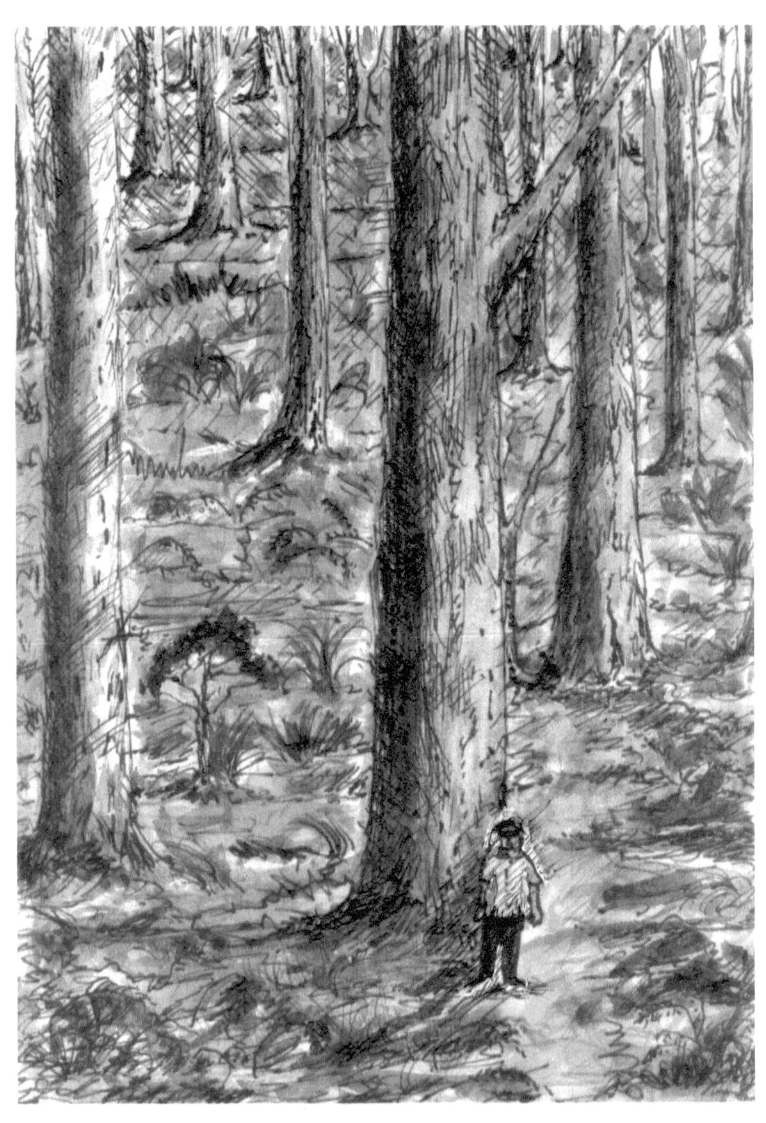

Karri Trees in a Valley

LOOKING FOR MARIANNE NORTH*

Karri (*Eucalyptus diversicolor*)

this grotesque burl absorbed
        you, but obsession came easily,
    propelled you to Japan, Sicily, Borneo

your portrait from Ceylon—
        a shawl-wrapped saint, cherubic
    aura and flushed cheeks framed
            by palm fronds

more restless and further
        flung you became after
    your father passed away

            your love transposed
                    to pitcher-plants,
                tree-ferns, bael fruits

now the same dirt track you took
        from Pemberton jars me,
    corrugations made more extreme
            by four-wheel tyre tread

than your trundling by horse-drawn
        wagon: easel, oils and implements
    of exile in tow to paint an aberration—
            a bearded gargoyle interrupting

pure ascension
your rendering is faithful
to forest spirits and to banksias grasping
with light-thirsty fingertips

by midday, septa
partition growth into gnomish faces, perfumed
bark below and above
shreds off in the sear

peering up unexpectedly—*Creation of Adam*,
Michelangelo in crevices, or, at least, seraphs
and sprites cavorting

you and I limber to the shaded
moss-strewn side   close our eyes   cling to cool
resilience, which is a karri
which is a tree.

**Marianne North (1830–1890) was an English botanical illustrator who travelled to Southwest Australia in 1880.*

SANGUIS ORTUS NATIVITAS

Red Gum (*Corymbia calophylla*)

I.
Once I saw blood,
I noticed it everywhere,
the glowing dark enamel

seeping from chambers
where organs pulse—
blood impregnating blood.

Wave after wave,
in the columnar light
of late afternoon, the tree

performs a plasma-letting.
I dodge feasting flies,
taste flecks of saccharine

disintegrating on my tongue,
imparting an acrid sting
agreeable as an antiseptic.

II.
There are lineages in blood—
bloodroot spicing the bland wild
roots of King George Sound,

bloodroot white under oaks,
the profusion of my blood
after a summer camp slashing

how it spilled like a springtide,
an open spigot in my eyes—
I thought it would never cease,

strange spangles of magenta
along a suburban Seattle street
after the drunken night ranting

of my neighbours, my laconic friend's
vessels ruptured from the altitude
of the New England Green Mountains

(his only ebullition of the day)
my scalp sopping fresh red paint
after the surgeon excised a lesion.

III.
This tree remember the births you
never saw, including your own—
our blood is everywhere, though
our bodies dam it back in remotest
gorges, it gushes forward at these
unlikely moments of communion—

blood *ortus sanguis nativitas*
                    blood.

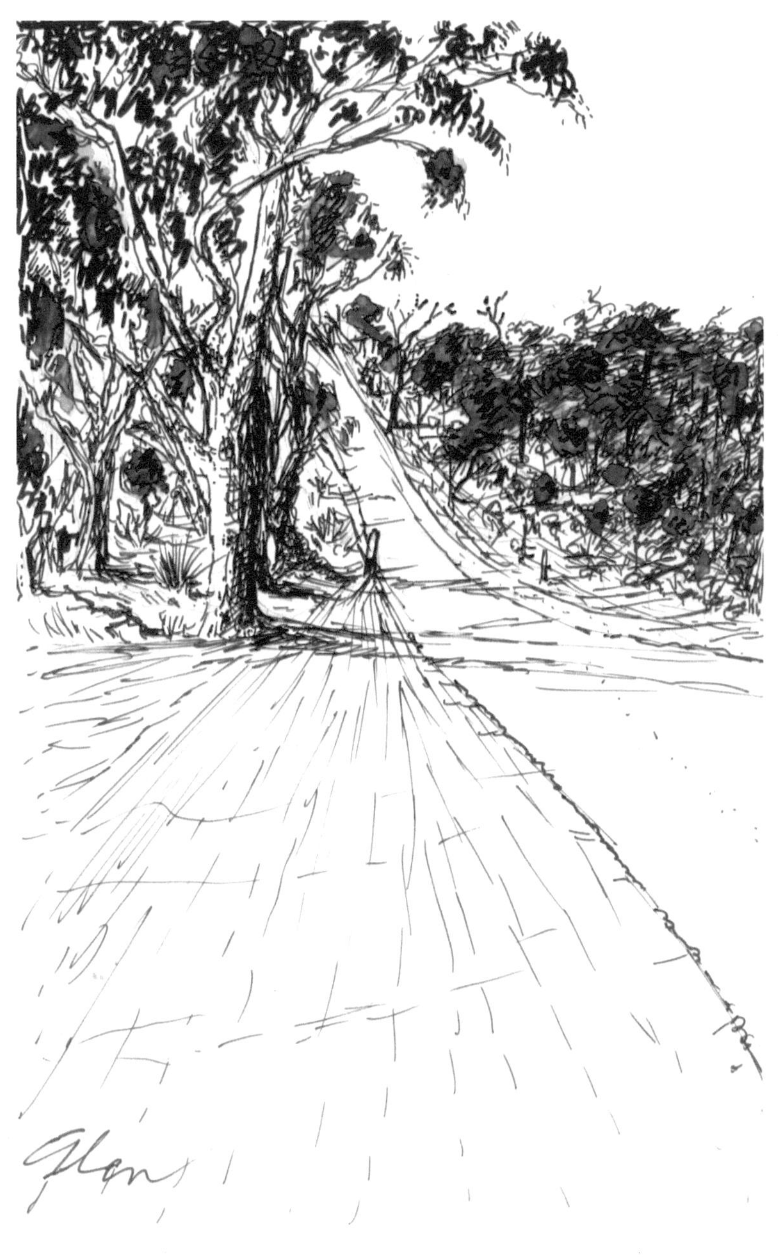

Salmon Gum Road

PHYSIS: OF AN OLD TUART

Tuart (*Eucalyptus gomphocephala*)

I.
its slow-growing trunk
ascends to arid heavens
carrying breath prints

II.
new faint golden tufts
under gargantuan tuarts
were hibbertia

III.
crags sequester rain,
like grey gutter water poised
to burst on daydreams

IV.
unhurried droplets
enlivened greatly to red—
coral vine petals

V.
swollen flower caps
have ruptured leaving only
stamens of moonlight.

THE UNSHROUDING

Red Flowering Gum (*Corymbia ficifolia*)

Deep River is low at the height
of summer, no tumbling of Fernbrook
Falls. I pack the boot swiftly and rattle
four kilometres on an unsealed road
to the South West Highway, nearly empty
as it meanders down to Bunbury;

through the karri north of Walpole,
peripheral coruscations in the olive-hued
bush, constellations of the Albany gum,
denominations of early *bunuru*: honey bee hum
within magenta traceries. I stop but once,
bleary-eyed witness to the unshrouding.

I SAW THOREAU ON MOUNT CHUDALUP

Balga (*Xanthorrhoea preissii*)

Up the inselberg, or *monadnock*,
                    after Mount Monadnock,
          the place where Thoreau anticipated God.

But this antipodal granite outcrop
                    on a sclerophyllous plain
          admits the admittedly less intrepid

sojourners lumbering past
                    balgas' blackened figures, like statues,
          seeking previews of Windy Harbour,

encumbered by knapsacks   toddlers   tattoos
                    *un*encumbered by clothing,
          interjecting hoots and burbling

into a concerto of snottygobbles—
                    their walk pleasant no less
          yet without transcendental largesse.

ARBORETUS SYNDICUS

Karri (*Eucalyptus diversicolor*)

I.
United Nations of the arboreal
                    a collective   a syndicate
          transacting lateritic business
at Big Brook Arboretum
                    copper-barked redwood
          same *Sequoia* as the Chandelier
Drive-Thru Tree in California
                    wispy boughs of evergreen
          Pinus waxy leaves of
                              turpentine *Syncarpia*
a different *Pinus* of Spanish extraction beside
                    Mexican cypress *Taxodium*
all intermingling with *diversicolor* candles:
          together we unfold pastelly upward.

II.
arching back to observe hypnotic
                    heights   bustling birds
          I discern a grey-bearded gentleman
statuesque on bench, blending with foliage
                    as if behind a hunting blind
          good thing I resisted urges
to micturate in restful arboretus ambience!
                    exchanging nods and half-smiles
          passing in unspoken agreement
                              I leave the man to his immense
meditations on time   history   nature or
                    the bedlam, which is our society
and resume my feline behaviour
          in another sector for propriety.

AFTER VISITING GUY FAWKES NATIONAL PARK

Beadle's Grevillea (*Grevillea beadleana*)

The bus, weighted
With conservation students,
Bogged in the greasy autumn
Mud of the road zigzagging out of
Guy Fawkes River National Park.

Darkness dropped
Like a mallet around us. The students,
Too afraid to alight. The rest of
Us huffed and hacked up some downed
Branches to lend enough traction for

The beast to climb.
We had come there to survey
Rare Beadle's grevillea flowers, their
One-sided mauve racemes, upright
As blood-hued horse-brush bristles.

Once presumed extinct,
(The species, not the passengers)
They were rediscovered in the '70s
And now are known to populate a mere
Five locales in northern New South Wales,

Much like the one we visited:
A sanctum of ravine-crossed country
Pollinated by eastern spinebills,
Yellow-tufted honeyeaters,
Crimson rosellas and the less frequent

Undergrad feathered in fluoro ribbon.
With their silky deep-lobed penmanship,
The Beadles resembled bonsai among
Less mature sheoak-wisps of the friable
Slope. Their red flowers scripted a fusillade

In the thick olive-green bush.
The bus snarled up the gradient. Its pallid
Light frenzied spectral shapes into motion
But, by the time we reached drowsy Ebor,
The fusillade had softened into an afterglow.

SHEOAK REVERIE

Western Sheoak (*Allocasuarina fraseriana*)

Welshpool Road mounting
the Hills above Perth City
soused in eventide spawl

sheoaks rummy with radiata pine
all fogging together—
my mobile reverie cut short

by the uphill travail of three
cylinders. I lapse into a mindset
beveled into white or red oaks

ejecting lacquered acorns
to the boreal duff somewhere
on a tenebrous broadleaf floor;

lore hunts us down the same
Nantosuelta lurking on the plain
feminine oak or the settler's bane;

tiny teeth are your verdure
neither as leaves nor as needles
but as cladodes, unlike the pine,

you see, where I come from
winter is roughshod and slaps
the rubicund faces of boys and trees

threadbare smiles crack coldly,
fall in brittle leaf potshards,
marrow hardens and turns icy

and just when you get used to it
the thaw surges in overnight
I know, this is somewhere else

but further along, at the roundabout
the bald tyres of the Daewoo skid
on the slick bitumen to Kalamunda

and I end up facing backwards again.

TO SPRING HOPE, HERE AND THERE

Golden Grevillea (*Grevillea pteridifolia*) et al.

"Hope is the thing with feathers" – Emily Dickinson

This morning, a kookaburra surfed the fence post
outside my kitchen window, headstrong in the squall,
unfussed in the gale, a stubborn figure in a snow globe.

Furry green apricots plunged to the ground, sprawled
in litterfall, pirouetted into the parsley patch (yellowing
and gone to seed) pinged the tin roof as noisy miners

mimed in feather bushes nearby. But, for you, Christmas
trees will soon bloom in fiery coronas bellowed by falsettos
of long-billed corellas, and kangaroo paws will overstep

the ancient scarp. I remember the sensation of noonlight
sharpening to summer's taper on my neck near the river
of the black swan, as sulphur-crested cockatoos wheeled

daredevilishly in a mob on the horizon. Here, crimson
rosellas veer then disappear into the olive orchard, hulking
wallaroos chew wadges of couch grass, magpies gargle

among their parliament before rain gurgles into the concrete
tank—and when I think I understand what ensouls us all, this land,
it bolts from the warm inside of my hand—green elusive pulse

of wren and wagtail, our filigreed dreams of hope unleashed.
I won't forget, at Jarrahdale, when ring-neck parrots seesawed
in the golden grevillea and, at dusk, red-tailed black cockatoos

rummaged for marri nuts with their beaks hard as steel pincers.
I think hope is as tough as feathers—a seed scatters, a bird shudders
on spectral wings, sings of things we must have faith in to imagine.

II. Tree Tales

Glen Phillips

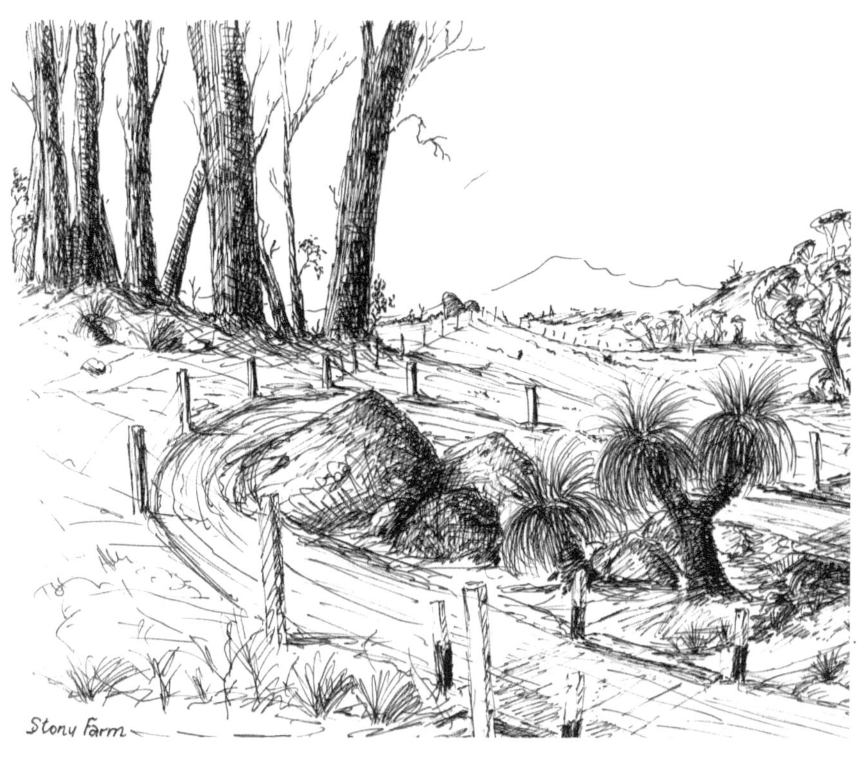

Stony Farm

DEATH OF A SHEOAK

Western Sheoak (*Allocasuarina fraseriana*)

Abundance comes suddenly
to our sheoak standing tall
these years in the bush garden;

for the burred sheoak seeds
I had played with as a child
on the sandy road to the farm
of my grandfather were falling
now in a dark rain.

Daily I swept them away
briskly from the brick path
where underfoot they sprawled.

That child who gathered them
so long ago, making his own
miniature farms of the strange
spined nuts, counting
his black sheep flocks,
now pauses broom in hand
at the far edge of life's span.

They say a tree senses
its imminent decease
and will profusely flower,
litter its seed in the face
of final fall from grace.

And sure enough, these days
suddenly our sheoak turns
from green; its needles
drooping pale—whole tree
sallow and sere. I am sad
its familiar place in our days
will be gone.
                    In our garden,
perhaps, this ageing tree is
the first to leave us, as my
own grandfather did. Others
too will follow one by one.
                    It goes on.

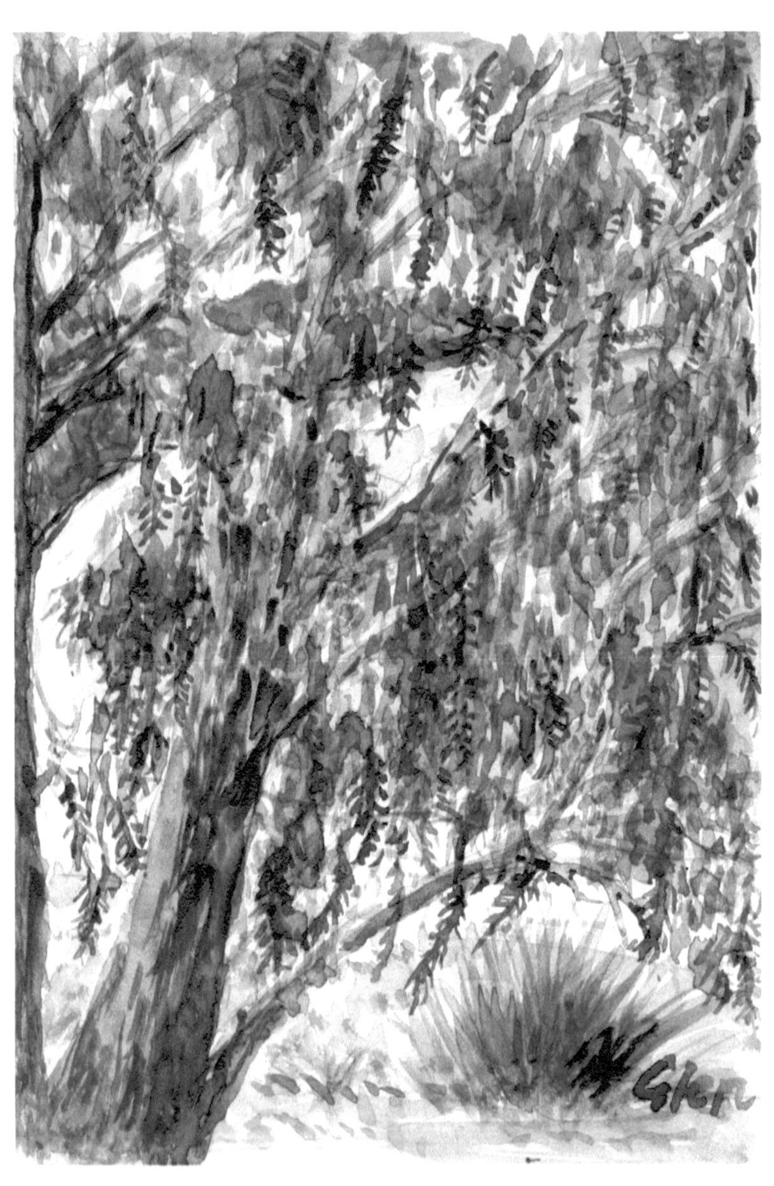

Coastal Peppermints

COASTAL PEPPERMINTS

Willow Myrtle (*Agonis flexuosa*)

Deep in peppermint groves we would scale
our morning catches—bream, gardies, even
a prized kingfish— spread out on rock. Stir coals
of the campfire ready for the grill. Then
lean back against a rough trunk to try taste
of our new bush cooking. At night we squeezed
on camp beds in the open, as sea winds
sighed in the drooped peppermint canopies.

Those summer holidays by the sea brought
multitudinous glimpses of the deep
blue of the Indian Ocean slashed by foam.
Seen through boughs of these peppermints, steeply
moving waves restlessly denied inland
steadiness of strait-laced wheatfields and tree
trailed roads. Weeping *agonis* boughs bore scent
of the pungent tropics mixed with the free
tang of sand-dune heaths. Darker angst too
of beached seaweed strakes belched up on the strand.

Maybe strange passions moved in us, prompted
by these groves to kneel, loving moonlit land?

ACKNOWLEDGMENT OF JAM TREE COUNTRY

Jam Tree (*Acacia acuminata*)

> *"The wattle seed in its curled throw-out pod*
> *is the sound of ejection in the heat . . ."*
> —John Kinsella, "Some Sounds of Jam Tree Gully"

Albert Alfred Wood, our grandfather, settler on
Yenyening Lakes' shores, when I was a four-year-old.
told me jam trees make best fence posts in the land.
True or not you're still a tough little *acacia*
acuminata. He'd also have known you as *mangart*
or 'spear wattle' or his name shouldn't have been Wood.

Admired by generations of *Ballardong* people,
this tree of blood and butter bi-coloured heartwood
appealed to Grandad. He tempered hoop-iron
strips into sharp blades and fashioned his hand
knives with jam-wood handles. Handsome,
they were his trophies lined up on the workbench.

But the jam tree deserves to thrive secure
in stands along road verges, unmolested
in bush reserves flowering in *Kambarang*
to make gold shadows spread at its feet.
And surprised settlers from afar slicing open
a branch for the rich odour of raspberry jam.

To the *Ballardongs* this *mangart* was
prized for exuded *menna,* the sweet gum
oozing in great gouts from branch and bole
in the *Birak* summer months. Their bounty
sweet and chewy, ('bush chocolate' some said)
but better bounty for spears or digging sticks.

Back to my first four years and wheatbelt jam trees;
standing by salt lakes they also gave me *menna*
to break off and chew until youthful jaws ached.

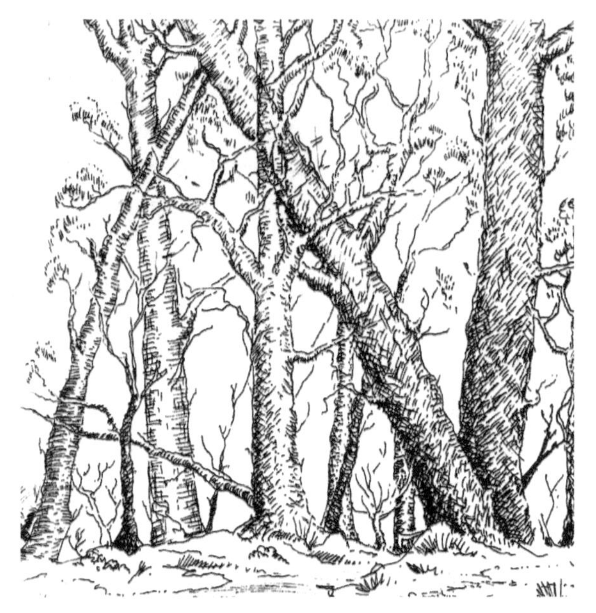

Red Gum Forest

MARRI MARRI

Red Gum (*Corymbia calophylla*)

Oh marri, marri, ever the bridesmaid
to enduring jarrah, that rainforest queen,
you're always flashing sweet creamy blossoms
in seeming wrong season, breasting wounds
like a fresh messiah or warrior with hard-won
battle lesions. Gouts of it or mere smears
for Noongars to gather as red-gum gold,
useful to weld haft to flinty blade of axe
for cutting footholds or to brew in hot
water for bush medicinal draughts. In
its leaf marri is ragged, unlike refined
margins of the narrow jarrah leaves. And
your bark is worse too, gnarled and dark. Where
jarrah runs vertical in straight lines elegantly.
There should be songs for both trees—jarrah-me,
jarrah-me; marri-me, marri me please. Only
in fruit is their real contrast: tallest is tiniest
in its eucalypt buds while marri sprouts
sprays of great gumnuts, green bobbins
as fit feasts for black cockatoos. Red-tails
matching pure fire of the redgum's bleeds.

BACK TO SWEEPING THE PATHS

Hakea spp. et al.

> *"I have swept away your sins like a cloud. I have scattered your offenses like the morning mist.*
> —Isaiah 44:22

Still I sweep brick by brick the paths here,
between hakeas, acacias and red gums.
Nevertheless, the age-old flourish asserts
from left to right with swish and swat
so dust motes fly and leaf and twig alike
scatter in gutter and garden bed there.

Brick after brick, row after row, our deft
sweeping thrusts bestow a tabula rasa
to this stairway to my stars. The whisk
is worse than the trace of rolling tyres;
for they merely batter flat these felled leaves,
grind odd gumnuts in crevices of bricks.

So then, back to sweeping paths we go fro
and to in this age-old art. Our bodies
do their twists in servitude to the brooms—
deference to peaceful Jains who precede
each footfall with the same flick of the brush
as when Curlers on ice assist their stones.

Thus we ease our sins with sweeping, spin out
our days with diligence on morning bricks—
sweeping is penance and protestation both.

MRS. DANCE'S JARRAH TREE

Jarrah (*Eucalyptus marginata*)

Why was she chosen to take
up the axe to the tree to make
sacrilege of someone else's land,
this Wadjuk owned sandy strand
by the blue Swan River, planned
capital of a new British band?

Contemplating that jarrah bole
with its rough bark, her role
as axe-person might have led
her to conclude instead
the main reason for to climb
that sandy bank was, at the time
she was the only colonial female soul
not pregnant in an Arcadian hell hole!

ROTTNEST TEA TREE*

Melaleuca lanceolata

After our dunking all day in clear seas
of Wadjemup's rock-bound bays, we glimpse
sun is setting behind these paperbarks.

The fine cupola fringes are their
crowning foliage. Such a hard won
right to survive millennia among

limestone ridges, marsh and dune sands.
Called 'tea trees' because of the leaves,
lance-shaped, etching now a fading sky.

Black paperbarks they call them; or
moonah to local Noongar tribes, though
shelter now for holiday weekenders.

So many eyes at evening have peered
at these dark outlined secants. Even
the Wadjuk prisoners under guard.

Dark eyes beholding dense pall
of melaleucas' leaf litany overhead
might also turn to glimpse with grief

that dim blue coast to the east. Trees
here pre-date human footprints, yet
share history with scales and fur

and feathered wings. Rottnest of
tea-trees dismissed by de Vlamingh
as a 'rat's nest' should teach us,
a humility we may well lack.

**Rottnest Island, off the coast of Perth, was for decades a prison for Aboriginal people.*

PEPPER TREE FARM

Peruvian Pepper (*Schinus molle*)

Outside the 'old house'
opposite the new farm home
a limb creaked all night on
the iron roof. That pepper tree
has re-rooted in my memory,
created its suckers sprouting there
connecting back to the dark
nodulous trunk that we walked
around. Found places on the dirt
swept clean by its acrid fronds.
Playing, we tasted peppercorns
but spat them out in pained disgust.

Lining the sand track to farmgate
were more weeping pepper trees.
Perhaps my grandfather viewed Peru
as appropriate to provide verdancy
to his 'Woodlands Estate'? The rest
were grey-green morts and tawny
casuarinas or banksias. Odd semi-
parasitic sandalwood or quandong
offered a greener shade of pale
but the pepper tree glade was settler
statement perhaps of interloper's right
to mark Jdewat land with foreign trees!

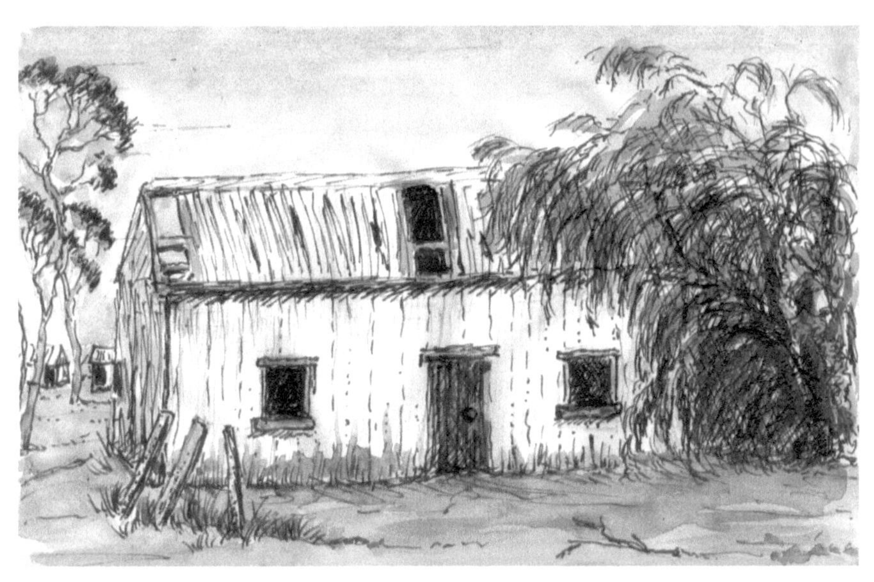

Pepper Tree & Old House

IN DEFENCE OF TUARTS

Tuart (*Eucalyptus gomphocephala*)

First bulldozers came, then mounted police
on their white chargers opposing us. We strove
to defend a corner of Ludlow Forest for love
of *gomphocephala*, white-wood giants, twice

the age of even jarrah trees. Named from
their club-headed nuts, these eucalypts were
remnant of great tuart forests—once from here
to the Swan River and beyond. We'd come

down from the Swan to protest clearing these
forest margins, already thinned sometime
in the 1830s then to ship as prime
plankwood to England or nearer Indies.

These banks of the Sabine River did not
need new rapine of what *Wardandi* called
'dou-art' trees. So descendants are now appalled
seeing *moondong** gathered at this spot.

Tuarts were known pre-invasion, settlement
or whatever you call it. Leschenault,
in 1802 sighted 'tuarts.' As a result
their Latin scientific name's now ancient.

And it's the tree matters. Elongated
crescent leaves, trunk like the soaring bone
of some megafauna's limbs rooted in limestone
and sand—tuart roots seem strongly related.

In their last March flowering, seed pods tore
off shy-hatted green domes before each bloom
arrayed its white shower in arched green gloom.
For that we challenged axe, saw and the law.

* Moondong *is the word for the white man among the Aboriginal Noongar people of Western Australia.*

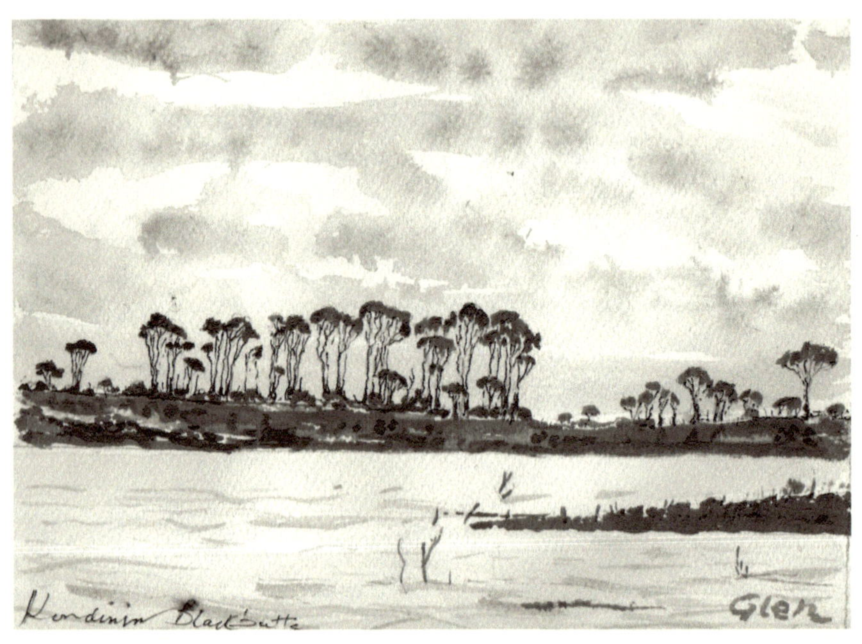

Kondinin Blackbutts on Shore of Lake Kondinin

FOOTSTEPS AMONG TREES

(a salty sestina)

Eucalyptus spp. et al.

Dust pinches between bare toes as you walk
red soil bushland paths, sensing bud and leaf
underfoot. It could tell where you've sprung from.
At five years, eight decades is your life-span,
an unknown destiny still left to run.
And yet the path curves on in shadow or sun.

Stroll on, youthful traveller, taste dust and sun,
breathe in scent of eucalypts as you walk,
savour tinctures of heath or wild broom. Run
your hand over rough bark, finger fine leaf
of wattle or smooth limb of blackbutt—span
of your hand to grasp. Here's where you've come from.

Down at the lake's brink among samphire, from
the lapping waters mirrored by the sun,
come low notes of water birds a wingspan
away it seems: ducks, coots, black swans. You walk
further, see swamp gums and banksias in leaf,
prints on the beach where waterhens have run.

At mother's call, turned from the shore, you run
towards the shuttered home. There waving from
the verandah a figure, frail as a leaf,
summons you now, 'Come in out of the sun!'
Yet still to-and-fro at margins you walk,
shirk being where Adam toiled and Eve span.

At kitchen table you contemplate brief span
of time you have come—and days left to run
before prison house of school bids you walk
forward, desk to blackboard, to perform. From
now, authority's shadow hides your sun.
Yet through window light you still glimpse a leaf.

And from the book of youth you take that leaf,
print with your feet the way of the brain's span
immense; then with the setting of the sun
unleash your fresh young mind and let it run
its imaginings as your footsteps trace from
your place of birth as far as you can walk.

You should cherish walking out, every leaf
at your feet fallen from heat of the sun,
unheeding such spans of time left to run.

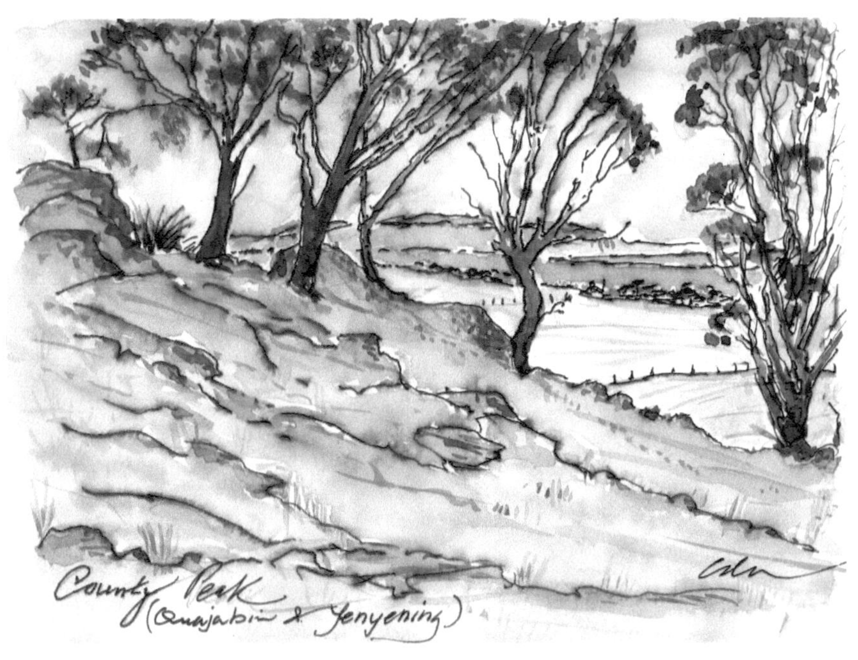

Eucalypts at Lake Yenyening

III. Tree Forms

John C. Ryan

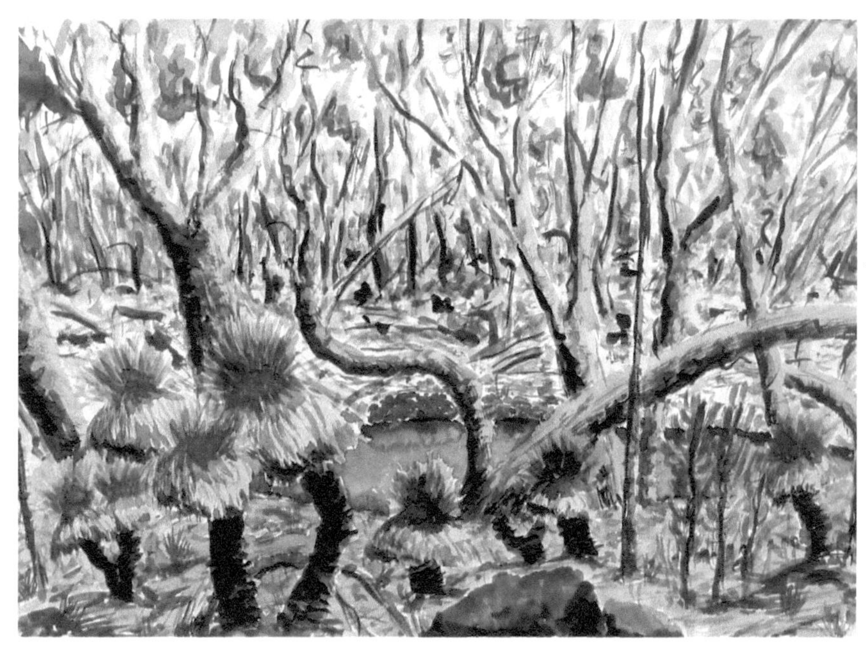

Jarrah and Marri Forest, Darling Range

THE CHURCHILL OF NEW ENGLAND

Port Jackson Fig (*Ficus rubiginosa*)

I am one of the fig trees at Mount Yarrowyck bearing
    An uncanny likeness to Winston Churchill, were he
        Reincarnated as a Green Man: a imperious chlorophyll
            Sourpuss with leafy jowls, bloated stomach, barrel chest

And generous buttocks. The wind has blown off
    My top hat and blown out my cigar, but I sharpen my
        Oratory in the presence of whoever will listen, for instance,
            An impressionable young wattle or new generation of duff

Eucalypts. And other audiences. But I cringe a bit at
    This crude comparison, for to blemish *Ficus* with politics
        Seems gratuitous and does no justice to the consummate
            Magnificence of my portly persona on my plinth of granite.

Doubtless, some bird once puked or pooped me out
    Here. And I deigned to deem this rock home-enough.
        My single root like a bonded-pair cable has intruded a
            Fissure and plunges whole-heartedly into dry *terra firma.*

Now I am free from the trauma of the Iron Curtain era.
    Yet my pomposity belies a sensitivity. An affection for the
        Brusque warmth of monoliths. A forbearance with wasp
            Larvae of my synconia: those flowers stinging me innerly.

FIGWARTS

Moreton Bay Fig (*Ficus macrophylla*)

"Figs yellow turning red, usually prominently warted"
—Flora of New South Wales

herd
of feral goats, snow
coats bolting upslope
or have they absconded
their paddock beyond
to a feast of fig
hope?
*

gum bark
stripping streamers,
tree unzipping trousers,
or was it billy rutting that
left these frilly jutting
splats of rusty red
around it?
*

blue-tongue
nudges head between
boardwalk planks to glean
bush flies zinging by, or is
she simply saying *hi* to
blue sky while it can
be seen?
*

fig tree
espaliered to granite,
splaying tentacles around it;

as cicadas call in counterpoint
I swivel on my ankle joint to
grasp the woody limb
that spans it.
*

latex like
milk exudes when
bark is wounded and then
from warted skin of fruits,
hard and green beans in
groups between rust
stems.
*

flowers
turn shyly inward,
inflorescence splintered
into ovaries translucent, juicy
as vesicles of ruby grapefruit,
wasp-churned through
this winter.
*

banyan
lichened to blue-grey
by boulder it pours over,
a lithophyte, stone-lover,
suckering up top—who
dropped it there, a
rosella?
*

this
fig is shapeshifter,
polymorphous stonelifter,
creeping body cables through
slimmest creases—root lace
to stone eyelet, or, likewise,
seed to sifter?
*

this
fig is freestanding!
hmm, wait, maybe not: did
its seed vessel take a different
landing, slip off its stony loft
into rubble wedge it's now
commanding?
*

fig
nested in crook
of gum who mistook
a feather-lifted fruit for
a casual visitor wanting
just a one-night
nook.
*

ficuses
in fields growing
huge, showing sculpted
muscles, flexing six packs,
ripped lats, perfect pecs
hot as molten lava
flowing.

*

a
currawong,
darts among upper
branches, taking chances
with sudden lances
of its beak and
tongue.
*

BICENTENNIAL TREE, MIDSUMMER

Karri (*Eucalyptus diversicolor*)

this Warren River morning

never quantifies chill and chatter
never divvies out bliss in numbers

the river's passage languid and

utterly nonmathematical as I sway
retrograde above for perspective on

this 200-year-old occupation

marked by 165 steel rods drilled
into the lustrous corpus of a karri

its 246-foot vertical plummet

from high to low negotiated
by 7 stout tourists donning sandals

how do they do that? I mean

in flip-flops with pegs through decrees?
sometimes things don't add up.

SINGEING RADIANCE, DANGAR'S FALLS, ARMIDALE

Ingram's Wattle (*Acacia ingramii*)

<table>
<tr><td>in full abandon</td><td>flowering</td><td>Acacia</td></tr>
<tr><td>ingramii at Dan-</td><td>gar's Falls</td><td>bursting</td></tr>
<tr><td>lucid aureate</td><td>pom-poms</td><td>seducing</td></tr>
<tr><td>bees with elixir</td><td>of early sun</td><td>springing</td></tr>
<tr><td>forward to witness</td><td>dangling haloes</td><td>lazing</td></tr>
<tr><td>over glorious brim</td><td>of vertiginous</td><td>plunging</td></tr>
<tr><td>to Salisbury Waters</td><td>underneath</td><td>cartwheeling</td></tr>
<tr><td>wattles gilded</td><td>are adroitly</td><td>acquiring</td></tr>
<tr><td>fire language</td><td>are combusting</td><td>chasming</td></tr>
<tr><td>with quiet sing-</td><td>eing radiance</td><td>consuming</td></tr>
<tr><td>swallows flitting near</td><td>blossoms-ever</td><td>goldening</td></tr>
<tr><td>head of falls</td><td>honeyeaters</td><td>trilling</td></tr>
<tr><td>as eels migrate</td><td>to far-off oceans</td><td>multiplying</td></tr>
<tr><td>inmost essence</td><td>of gorge-glowing</td><td>full in abandon.</td></tr>
</table>

ROCK ORCHID HYPHAE

Sydney Rock Orchid (*Dendrobium speciosum*)

Cutlass-shaped
leaf, rigid sandpaper sheet
smoothed from use,

but with gritty aftertouch.
Margin and midrib
surprisingly resistant when

strummed between
thumb, index and middle
finger. From tip

to base, faintly traceable
veins break out
in browning blemishes.

Profound gouges
found on hide-leather edge
where beetle mandibles

chewed abscesses—
charred blotches with rimes
of ash, like cigarette

burns on old mattresses.
Fitful wind shakes
organs of rock orchid—

whole stiff gorgon quakes,
transmitting shivers
along ridges of stretched

stems, those pseudobulbs,
half-clothed in membrane,
feeling of filthy paper

lantern material left outside
over many winters,
disintegrating and peeling

back. Bulbs, at distance,
reminiscent of plump
asparagus spears—squashy

rotten, half-heartedly eaten,
forgotten in refrigerator
bottoms but, to touch them:

sensation speaks truth,
upends expectation—fleshy
antennae of lithophyte,

as dense as antique wooden
umbrella handles,
fists clenched around them

on some squally amble.
Between stalky assemblage,
shaking slightly on verge,

and lichen-splotched rock
surface—rootlets
sprawl in air, their merest

earthly medium there,
extract what nutriment they
can from odds and ends,

aggregated miscellany
of gum trees—lumpen, dry,
wavy, uncooked noodles,

springy to phalanx pressure,
sachet ripped open.
Another dendrobium holds

vigil overhead, suctioned
firmly to sharp pitch
of granite—miniature grove

of yellow palms leans to old
medusa below, getting
closer yearly by millimetres.

Things live by touch, live by
being touched—
thrive in *becoming* touched.

False lily, soilless at gulch
brink, miraculously, yes,
but savvily too—how things

must reach out continually,
across yawning hugeness,
wayfaring by yank of feeling

like hyphal filaments,
unseen, spindling through
inner orchid circuitry.

At Paradise Rock Lookout,
fringes of ecosystems
interbreed in ravine creases,

stone anatomies at horizon
are femoral heads
articulated with acetabular

rims, waiting, for millennia,
to stand upright,
stride off into opaque light,

across terrains of glacial
reminiscence—
landscapes echoed within

bodies roused by
feltness. Termite moundup,
conical adobe oven,

concrete-tough from sweat
and spittle epoxy
of billions of wood ants in

holy clearing of burnt
eucalypts, acrid with scent
of carbonised stumps

growing potent with sun,
ripening among young
sheoaks and mossy pendula

which insinuate rainforest.
Whiff of fire incites memory
incised in Apsley gorges,

limbic impressions of being
in touch, of beings
in touch—bunches of herbs

with downy peach fuzz
fragrant horehound, palmar
arches open in welcome.

Other shrubs bring to mind
rosemary but with-
out woody camphor aroma,

hemlock-like evergreen
needles pliable and yielding,
to wit, neither briery nor

wielding ordnance of any
kind. Jurisdiction
of king orchid, outstanding

dendrobium, imperfect
rock lily, not-yet in blossom
but soon-to-be, creamy

flowers about to awaken
synchronously,
scent glands over perianth

poised to perfuse stingless
bee-dazing
polyphony of boulder ledge

attar but, for now, there is
touch—most profound
and immediate of senses,

for Diderot. A skin-
knowing not always in flesh,
but of cuticles of beings

in communion nonetheless,
to stretch filaments
in airy possibility, to breach

chasmic spaces between—
threads of hyphae,
unseen reach to deep green.

THE GUARDIAN TREE

River Sheoak (*Casuarina cunninghamiana*)

Y Y
I I
am am
your con I
prot vec I
ector tor I
to live fore I
subside creek am
beside song lisp b
Boorolong wisp b
swind rush crisp b
sprigs hiss lapse b
thigh sighs ellipsis I b
rivery riffs garish but a b
cassowaryish lavish m b
pick of filaments li b
in bird hymnals. st b
bark skin splints en w b
into kookaburra ing w b h
red-winged parrot hat b h
& cockatoo scripts. way ut h
I've witnessed one will can o h
thousand eclipses wind you o h
glimpsed distance whisk hear o h
immense in senses today me? o h
of your reluctance away h
to confide internal in w
striving & desire for h
meaning with other. er

please bury your soul erer
in sodden sand below rere
& my roots shall rap haha
your mindstream & veveve
my seeds will release youyou
on samara wings beenbe
riparian sovereign I am
pollinated by windriff
fruits free by creekphrase
idioms fixed by nitrogen
fraternal with soil *Frankia*
sprays inside of electricity
intelligent circumlocutor
of floodzone & around
for eons & then some
but where've you
been & tell me
tell me who
are you
now
no
?

this earth you & I share is so much is spirit sustenance is you & me within & of us

I am y y
speak thought y y
through bafflegab y y
laughing belief me y d d
gravity is shocking y d g g g g
my master a flock of yy d g g
syllables sheoak textualityy d g g
a steward of pure words ddd g g

bulbous growth at baseeeee g   g
bifurcated trunk arching ggg
sphere warbled into beingg   n
by magpies see my insides   n
disclose blackening scars   n   g
of fervor I warp my skin   n   g
over medullary traumas   n   g
as I prudently don this   n   g
coarse coat every season   n   g
this is why I live by divul-   n   g   g   g   g
ging to whoever wants to nnn   g   g
pause, look, learn & listen nn   g   f   g
my lingua is not uncertain   g   f   g
but brandished graffiti-like g   f
sun soothe intensifyinggg   f   f   f
brings bitterest pleasures   f   f
my elders passed I at last   f   f
an island within an island   f   f
Boorolong hurrying past to   f
Gwydir whisk of mourning   f
off stupor & I awoke to find f
you slowing towards dark of f   m
mine carmine-heart swollen   m
rivers flowing from under M   m
Yarrowyck shoulders bearing   m
loneliness but to be denied this   m   m   m
is worse than your hands trem   m   m
bling with an expression tense   m   m
gazing into open gashes of me   m
don't leave yet don't forget that   m
I forgive you safeguard the banks   m
maintain an equilibrium inner m m m

river sheoak, giant river oak amm m
neither beefwood nor larch but m
similar enough to both not mere m
source of turnery yokes shingle m
but singular tree with opinion
russet phyllodes falling around
writing this text in presence of
wispy progenies at confluence
I am your guardian to live to
subside beside I am convect-
or fore creek song I am wisp
lisp lips crisp lapse of ellipsis
now tell me, where have you
been, who are you now, no?
and just where are you heading on this earth you share with me in spirit sustenance?

THE GUARDIAN TREE

River Sheoak *(Casuarina cunninghamiana)*

Y Y
I I
am am
your con I
prot vec I
ector tor I
to live fore I
subside creek am
beside song lisp b
Boorolong wisp b
swind rush crisp b
sprigs hiss lapse b
thigh sighs ellipsis I b
rivery riffs garish but a b
cassowaryish lavish m b
pick of filaments li b
in bird hymnals. st b
bark skin splints en w b
into kookaburra ing w b h
red-winged parrot hat b h
& cockatoo scripts. way ut h
I've witnessed one will can o h
thousand eclipses wind you o h
glimpsed distance whisk hear o h
immense in senses today me? o h
of your reluctance away h
to confide internal in w
striving & desire for h
meaning with other. et
please bury your soul erer
in sodden sand below rere
& my roots shall rap haha
your mindstream & veveve
my seeds will release youyou
on samara wings beenbe
riparian sovereign I am
pollinated by windriff
fruits free by creekphrase
idioms fixed by nitrogen
fraternal with soil *Frankia*
sprays inside of electricity
intelligent circumlocutor
of floodzone & around
for cons & then some
but where've you
been & tell me
tell me who
are you
now
no
?

this earth you & I share is so much is spirit sustenance is you & me within & of us

I am y y
speak thought y y
through batflegab y y
laughing belief me y d d
gravity is shocking y d g g g g
my master a flock of yy d g g
syllables sheoak textualityy d g g
a steward of pure words ddd g g
bulbous growth at baseeeee g g
bifurcated trunk arching ggg
sphere warbled into beingg n
by magpies see my insides n
disclose blackening scars n g
of fervor I warp my skin n g
over medullary traumas n g
as I prudently don this n g
coarse coat every season n g
this is why I live by divul- n g g g g
ging to whoever wants to nnn g g
pause, look, learn & listen nn g f g
my lingua is not uncertain g f g
but brandished graffiti-like g f
sun soothe intensifyinggg f f f
brings bitterest pleasures f f
my elders passed I at last f f
an island within an island f f
Boorolong hurrying past to f
Gwydir whisk of mourning f
off stupor & I awoke to find f
you slowing towards dark of f m
mine carmine-heart swollen m
rivers flowing from under M m
Yarrowyck shoulders bearing m
loneliness but to be denied this m m m
is worse than your hands trem m m
bling with an expression tense m m
gazing into open gashes of me m
don't leave yet don't forget that m
I forgive you safeguard the banks m
maintain an equilibrium inner m m m
river sheoak, giant river oak amm m
neither beefwood nor larch but m
similar enough to both not mere m
source of turnery yokes shingle m
but singular tree with opinion
russet phyllodes falling around
writing this text in presence of
.wispy progenies at confluence
I am your guardian to live to
subside beside I am convect-
or fore creek song I am wisp
lisp lips crisp lapse of ellipsis
now tell me, where have you
been, who are you now, no?
and just where are you heading on this earth you share with me in spirit sustenance?

I TURNED THE CORNER AND ENTERED THE MIND OF THE FOREST

Antarctic Beech (*Lophozonia moorei*)

I turned the corner and then I entered the mind
Of the beech forest. The seen was not a scene
But a psyche. The trees' old way of thinking
Coppiced from within me. I walked inwardly
A while towards eternity. It was no ordinary
Overcast midday before Labour Day. Should-
Ered by the Great Escarpment, I gaped east over
Spinal ridges of the Bellinger River Valley. I heard
The drawled and well-treed clauses of glacial speech.
Through haziness beneath, prone figures of Cenozoic
History sprawled towards the Tasman Sea: sacral
Curves, lumbar hollows, those vertebral foramen
Of time itself ever so expansive in its brevity.
My body dropped through basalt strata of
Other epochs as I rounded the elbow below
Point Lookout and crashed face-first into the
Very thought of the forest. Away from picnicky
Clamour. Farther away from the yowl and yammer
Of randy roisterers, of backpacking boisterers to a
Lyrebird percussing in the brush downslope from
Us. When I had to rush rudely by a camera-laden
Cadre of eco-tourists, sidestepping their hanker
For communion with the wildness of New England
And so leaving my feathery ground-dwelling fellow
To his flirtatious spring swaggering. Did I mention?
It was the day before Labour Day and there were all the
Typical signs of a prodromal state: edema and irritation,
Contractions, perspiration and the vague indication of
Colostrum, for some of us, that is, and that was how

I entered the mind of the ferny *Lophozonia* forest.
A vestige species, identified first by William Carron
And W.A.B. Greaves along the Upper Clarence River.
After that, Charles Moore, in homage to Carron, called
The tree *Fagus carroni*. Then Ferdinand von Mueller, in
Homage to Moore, renamed it *Fagus moorei,* though,
Before all of them, Carl Ludwig Blume propounded
The term *Nothofagus* for "false beech" but meant
Notofagus for "Southern beech." And so it was:
Lophozonia moorei, on pre-Labour Day, with its
"H" intact nonetheless despite agitations of genus.
Barbecuers bellylaughed at the comedies of treeness.
From the second lookout, I heard utes growl in first and
Second gears to Waterfall Way, everyone, including the
Forest, ecstatically indifferent to the accumulating "h"s.
I concur with Maiden: "I have quite satisfied myself
That the separation of *Nothofagus* from *Fagus* is
Justifiable." And Fagus, the Northern Hemispheric
Beech, a child of the Middle Eocene, a meagre fifty-
Million years it's been. But *Nothofagus* pollen can be
Seen in Tertiary sediments eighty-million years in age. A
Gondwanan taxon with recollection of supercontinental
Drift. Its bones ground in the rift between Australia and
Antarctica in the Late Cretaceous. It witnessed the
Era of Mammals. Then the trees witnessed us.
Although we name them, we cannot know them:
Red Beech, True Beech, Colonial Beech or Mountain
Beech, Negro-head Beech "owing to the rich dark colour
Of the foliage," Maiden noted. But, for its Indigenous one, he
Knew "of none although it is probable they had a name for
So conspicuous a tree." I turned the corner and entered
The mind of the forest. The seen was not a scene

But a sensation. The trees' old way of seeing
Bore winged seeds within my being.
so *lophozonia moorei* formerly
nothofagus moorei, speaking,
twisting, glacial beeches
along eagles nest track
populating escarpment
at juncture where yellow
asters, purplish solanum
and creamy paper daisies
are beginning to fade away
where acid of oligotrophic
soil is summoning raucous
congregation of epiphytes
mosses, ferns and orchids
along footpath girded on its
downward aspect by smooth
steel handrails and tidied up,
anticipating spring arches,
where a collapsed beech,
chainsawed, is disclosing
its clotted crimson heart
in coronary rays incised on
a cross-section of memory,
evanescent opaque views
over gullies made of gums
and wallabies "where the
land is frequently covered
in mist" as young botanist
G.N. Baur said in the 1950s
and when, nearing weeping
rock, knobbly beech shapes,

announced themselves as
caespitose, stunted, multi-
stemmed, tufted presentations
and clusters, gnarled limbs
burgeoning from boulders,
whole boles cloaked thickly
in lush assemblages of clingers,
generations of knotted trees
leaning in thronging synchrony
bivouacked to this scarp brink
scape fluctuating with ruptures
of water dribbling from bluffs
accruing in quagmires below
slick cliffs glistening in timid
sun, sudden microcosmos of
bracket polypore, undulating
undersides having the colour
of cooked salmon, and there!
oy, *dendrobium falcorostrum*
with succulent sectioned stems
tapering, though no porcelain
blossoms dangling yet from its
edge, exclusive to beech orchids,
sanctuary of banksia of platypus
valley lookout high up: could it be
lignotuberous *neoanglica,* its
leaves stippling in grey-white
feltness?

AT A BEND IN THE TRACK, I RETURN TO EARTH

Marri (*Corymbia calophylla*)

he could burrow to my pith,
his fingertips through my cambium,
once concrete-hard, now rotting
in sleeping boneyard of protuberances,
disfigured scapula, splintered sternum,
broken femur, heaped in middens.
down down daubs of defunct
termite clans, gangrene in my toes.
he kicks a clod, watches it roll down
slope, but high up me, stubby limbs bloat
like beached whales but without their
sick belches from under bleached skin.
all my colonisers have been stilled:
no red-tailed black cockies swarm,
no blood-splatter on begrimed bark.
washed-out palimpsests of me seethe
and even the sun shies away from
my slow leafless reckoning. I was a
giantess who slouched when mature,
who slouches now fading, acid red ants
hastening where my lungs aspire less.
I was once a behemoth inching heaven-
ward through inflections of trunk syntax,
towards numinous flecks of lumen.
he could spend the evening peeling
away my tetrahedronal bark layers,
sifting through these extinguished tints,
but farther along a kangaroo scratches,
stirs an earthbound tendril, whorls
an orchid leaflet, unfurls—I once
inhabited those worlds.

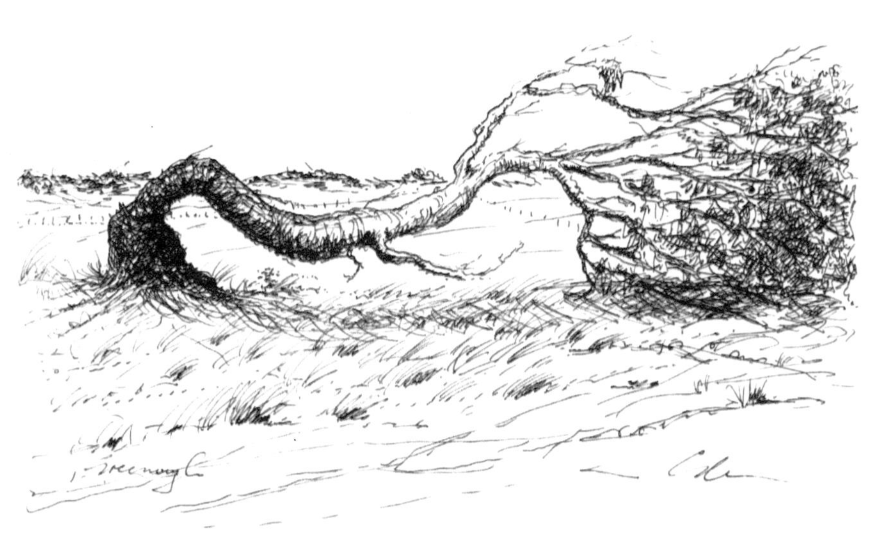

Leaning Red River Gum, Greenough

THE FUTURE

Wandoo (*Eucalyptus wandoo*)

in the photos of your weekend lunch   oat noodles float like
gnocchi in tannin broth   your mother plucks morsels of gristle with
chopsticks   son gnaws a dark barbecued beef bone   sea buckthorn
juice half-drunk   green straw an extravagant tendril twining out of
your glass of sunset nectar   *never worry, my dear one*

I will wait for you   with fragrant-flowered garlic
greens   dim sum baskets   chili tofu   sour cherry
jewels   little mugs of beer   amber   and steady   and
lazy   as the Yangtze through Nanjing   *your promise is*
very precious to me   more precious than any gift

here, the wandoo have dropped skirts   and gaze timorously
twist torsos like danseuses   sun-dyed thighs flash   ruddy buttocks
bulge   quadriceps flex   and ease   my objects of concupiscence
have been   these seasonal miracles   a nudist colony in neon
winter flame   a swallowing of the not-so Great Southern Highway

trending ironically east   15 ks from The Lakes
(that once were)   today the wind is kin to York rain –
petulant, inconstant, heavy   but to you this metaphor is
meaningless   *do you mind that I have a son?*
in the West, there aren't the stigmas you suffer   but
there are others

you will learn these vocabularies   just as I   learn how lichens
punctuate the shedded skin of wandoo   and will learn   where to
buy size 48 shoes in your city   and how to recognise the smell of
silk   your heart is clarion   and your loneliness is not profound
not pernicious like mine   nor has it ever been

a soft smile that is nuclear *on the street there are*
many beautiful women *everyone is beautiful* *every day*
is 32-35C your father, a kung fu master in Shanghai
wears prayer beads and a black shirt deftly steers a
paddle boat with three fingers melting ice cream
streams

down the stick you grasp like a fountain pen gushing vanilla ink
you gaze to your son your pink pants glistening at night I must
enunciate and listen closely otherwise rain pelting tin obscures
your sighs silences hesitations sudden lapses into
Mandarin you don't need my language to know I am not

a figment my pictures are of earth: spindly trees
of roos posturing like teenage weight-lifters, females in
summer heat with a tuft of face staring from
their pouches of ivory sand beaches Down South
cerulean seas butterflies lifted on leaf trampolines

but at Jinghai we will stroll Ming City Wall on Emperor's
concoction of lime and yellow soil glutinous rice and tung oil as
Nanjing irrupts like fire around us honey-comb apartment sliver
sleek bridge over Yellow River *I have been alone a long time*
now I see a beautiful future wave to me *this future.*

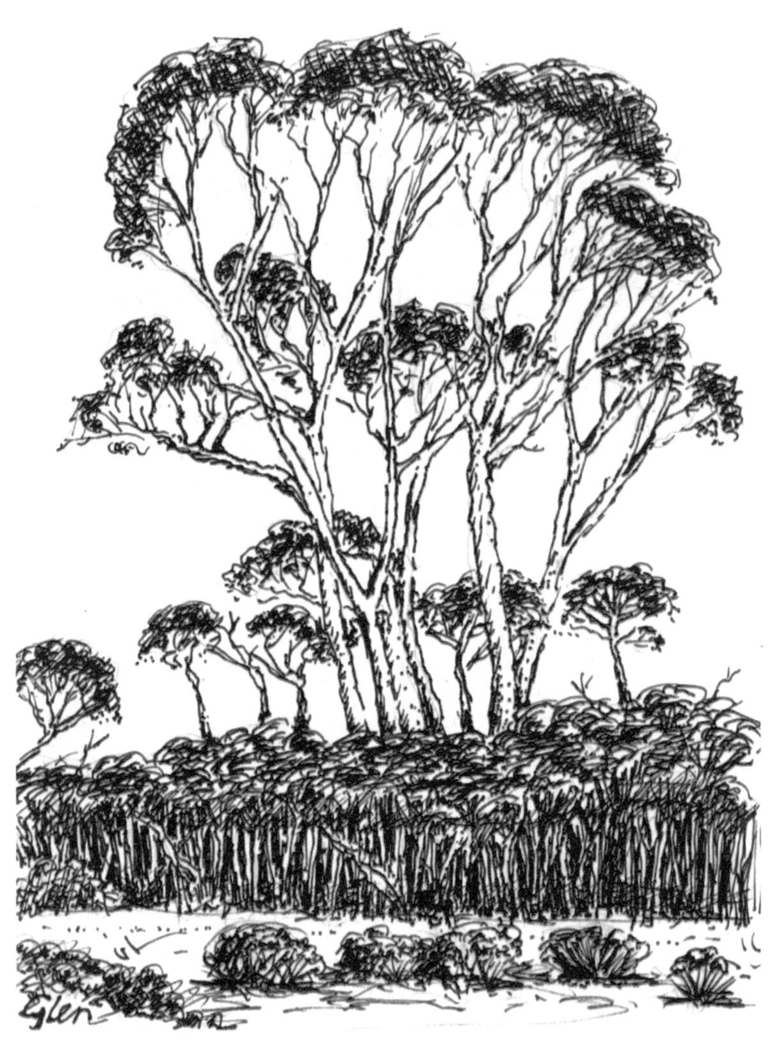

Salmon Gums and Wodjil

LAZARUS

Wollemi Pine (*Wollemia nobilis*)

I
am
pro-
tected
from theft in a
steel cage on a lawn
in plain view of kangaroos,
also behind metal fences. Leave me a-
lone with my two-hundred-million-year-
old sensations (!!) in ferny branches
that terminate in my father's seeds.
I am known as a Lazarus taxon for
a reason: deep valleys of memory.
I implore you to avoid them
at all costs because
there
are
dan-
gers
in re-
mem-
ber-
ing that
are far more perilous than anything possible
in this world. Your unimaginable brink is the one I hurl myself over.

IV. Trees Burn

Glen Phillips

FIREWOOD BANKSIA

Banksia menziesii

Seems one banksia at least needed to show
immigrants new ways of getting to know
how to make a briskly warming fire.
And Perth's settler-invaders with entire
families here in 1829
shivering in their flimsy little line
of tents and bush huts, thoroughly chilled out,
would soon be rubbing warmed hands no doubt.

So here on the Swan sandplain they learned
(perhaps from *Noongar** people) if they burned
the thick boles and boughs of such robust trees
as *menziesii* (or *grandis* too, 'bull banksia') these
would give winter warmth in times of trials
to follow. Woodcutters carved endless aisles
through scrub to seek banksias' rough
beauty. Few of the children played tough
with the huge woody banksia nuts that seemed
like monsters with many mouths. And screamed
when chased by bullies claiming to be 'the wild,
banksia men'! Applaud genius who compiled
famous storybooks of her gumnut babes, who
were bush fairies drawn from plants related to
*Whadjuk*** territory. It was Miss May
Gibbs from Perth who saw a whimsical way
with banksia nuts and ragged leaves to bring
to generations of children another enduring
existence for poor desecrated banksia groves.

So now I keep in jarrah-wood bowl the leaves
of that banksia and its gnarled nut. And show
children how much *banksia* they've yet to know.

**The Noongar people are the Indigenous owners of the Southwest corner of Western Australia.*

***The ancient area around Perth is place of the Whadjuk Noongar people.*

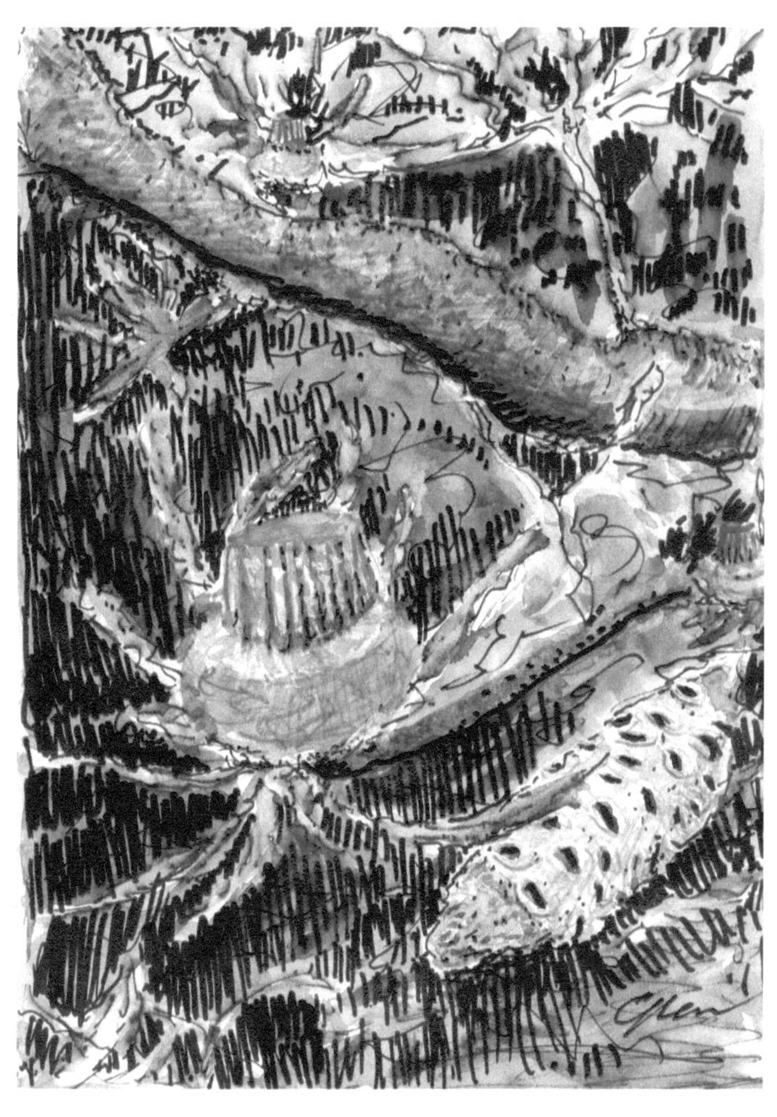

Firewood Banksia, Flower and Nut

CHARRED GROUND

(or, after an Australian bushfire)

Smell of burning's in the air—
xanthorrhoea's tortured trunks
are like limbs of the immolated;
while mallee saplings are
spent torches, shivering
their black tinfoil leaves.

And then the startled crack
of the still-burning heart
of a once-living casuarina.
Third degree charred ground
is scorched hairless of heath,
scarred with white-ash shadows
of what had been. A holocaust ago.

Dampen every ember's faint
signature to our conflagration's
past? We shutter behind eyelids
each lingering backward glance
to the fire we let consume us.
And yes, there's smell of burning
in the air. Burning of the air.

ETUDE FOR A TINGLE

(or Tree of Woman and Man, *Eucalyptus guilfoylei*)

Sure, outback salmon gums crane to the sky
in some places, blackbutts by the salt lakes,
white gums making aisles good enough for
country services along certain grave gravel
roads to kingdom come. But where were tall
hafts of forest karri, marri, tingle and jarrah?
Benevolent boles expanding ring after ring
of growth over hundreds of years, a thousand
even, is to defy gravity with their grace.

To every man and woman, it's honour enough
standing close to such grand cellular mass.
Our lifetime is brief interval for those we
traverse space with on our rocky sphere
spinning for a burning wanton sun among
fellow travelling orbiters. But heed now
the tremor at your back as you lean against
dark butt of this tingle's living column, feel
that faint stirring—as if it really breathes.

Our hearts are pumps but such trees raise
eighty metres or more of nutrient liquids
to sustain each bough, leaf and blossom.
Lift your chin and follow the smooth trail
of bark arcing up to clustered foliage.
This is roadway for ants and centipedes
or other fellow travellers: alpine climbers
like *kaarda* and *koomool*, lizard and possum.

Rainforests offer these many living spires.
And our
own small
towers of
vertebrae
only observe
attendant roles.

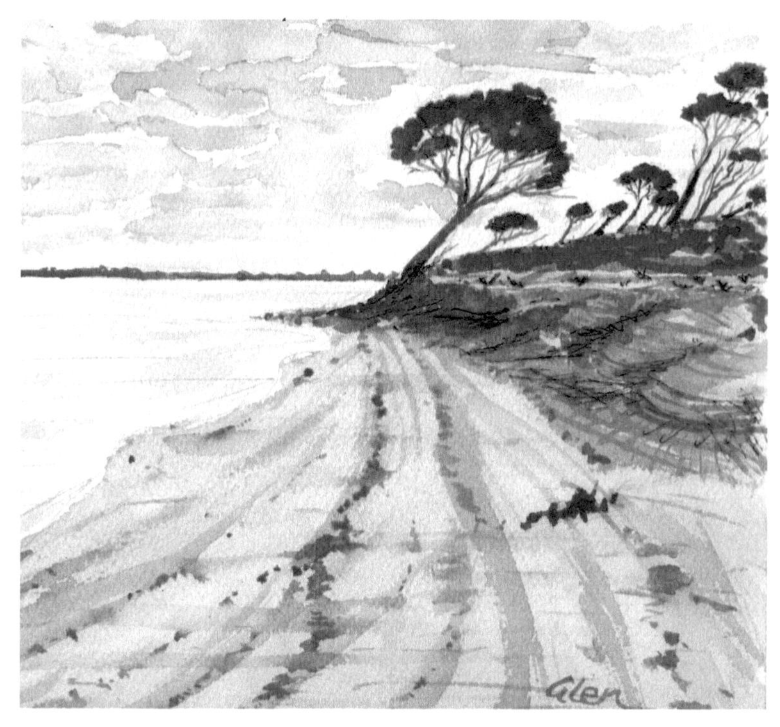

North Lake Grace with Blackbutts

FIRE TENDING

At countless fireplaces
I watched through cupped childhood hands
tending of fires
stick nests
for ignition

inserting paper wicks scrape of match
blaze we all desired

something like ancient wisdom
fighting dark matter
memory of millennial
hearths
when ice-ages attacked

or eruption frenzied blackened skies
month after month
grass stopped growing
fallen leaves littered

Searching for caves where they could
light fires
women brought wood
men hunted weakened prey

Once more I see   family hearth   mallee roots
          potatoes roasting
                    in embers,
                              making toast or
                                        toasting chestnuts

In the morning   only cold ash
          life sans-fire
                    of love
                              is without worth

We reached out hands   for warmth   last night
          it was no denial
                    of fire
                         in our care   consuming all

WHILE THE BUSH BURNS

(or, I'm Not Yet Ready to Go)

Still at high school when a weekend bushfire
raced across home paddocks. I looked up
watching from my open textbook from our
verandah as the fire truck and farmers' cars
careered to cut off the flames before the blaze
escaped into wooded hills. But exams were
the conflagration on my mind that day. And
next day, I recall, a schoolmate, farmer's son
seeing me at the station on the way
to school stopped me: "S'pose you were writin' poems
while we were puttin' out the fire. Or paintin'
some bloody watercolour of the scene, mate?"
The tone accusatory and I felt, perhaps,
just like those who were slow to enlist in
1915 when girls gave out white feathers.

On the train to school shamed, I curled up
by the window like some grey grasshopper
folded upon itself and thought: then what
sort of bloody poem would I write? What sort
of picture of a grass fire would I paint?

Now I recall ten thousand years ago
an ice age was here. And sure, that would have
put out the damn fires. And maybe drove men
then into caves to paint stories or poems
as in the Kimberley ranges. The nation
now calls it *our* heritage of Aboriginal art.
But of course it's *theirs*? That was a long time
ago, anyhow, for now we have governments

who first declare national parks and nature
reserves. And then slyly usher in mining men.

Another time—still long ago, but after my
boyhood failure to join fiery battlefronts,
I lived in suburbia near vast plantations
of imported pines. News came a murderer,
escaped, was lurking there with loaded gun
and a radio station trumpeted a call to men
(red-blooded ones with firearms) to proceed
at once to our street. And join the manhunt.
I hung my head at this, although neighbours
were eager to find more guns for husbands.
So as to lay their heads peacefully to bed.
Self-examining that night I was left
with no strong urge to put brush to canvas
or carve my distaste in stone calligraphy.
Ever the shirker was I? And still no poem?

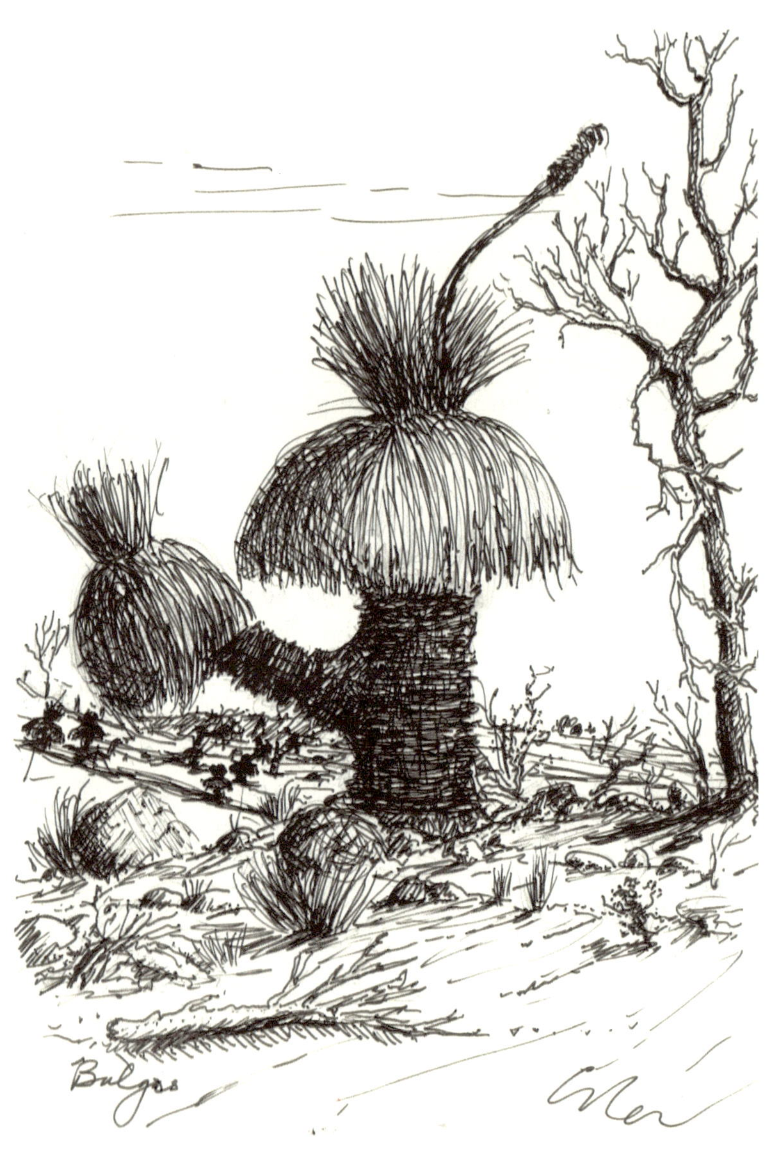

Grass Trees (Balga)

THE BALGA

Xanthorrhoea preissii

BB
AA
LL
GG
AA
Xa
nt
ho
rr
ho
ea
Ah, the balga! Seen in stubborn clumps with its green shower
of needling sprouts above a dowdy skirt of grey or brown.
This grass tree or xanthorrhoea is larder to other
life and rich provider of artefacts.
Driving through wheatbelt
country we could see them
marching across hill and vale
like armies of extra-terrestrial invaders. Robots
with their antennae aloft, their single foot burned black,
burred greybeards frowning through the forests and along river flats.
First I knew the balga on my grandpa's Yenyening farm when amber
chunks of its trunk lay in the woodbox with morning sticks ready to
get the great black stove alight in early mornings. How those
varnished chips of the fallen trunk blazed as our firelighters, filled
the farm kitchen with sweet resinous scent as grandma shifted pots
and pans from the hob, stoked on more whitegum shanks to feed
the crackling fire! Thankful we were, through childhood years, for
this free kindling and forever after that. It was our first gift from
Gondwanaland for us invading immigrants to this Great South Land.

But we were newcomers, for
Noongar people in many
millennia had learned all
and more than we knew.
That, for instance, the soft
growing core at the centre
could be good tucker. Or
that the gold gum from the
balga's trunk helped hold
flint axe heads firm on the
haft; fastened spear barbs
securely on whippy shafts.
But way before this small
birds nested in balga's beard
and beetle and grub found a
place to burrow in the trunk
for sustenance and a private life.

THE BALGA

Xanthorrhoea preissii

BB
AA
LL
GG
AA
Xa
nt
ho
rr
ho
ea
Ah, the balga! Seen in stubborn clumps with its green shower
of needling sprouts above a dowdy skirt of grey or brown.
This grass tree or xanthorrhoea is larder to other
life and rich provider of artefacts.
Driving through wheatbelt
country we could see them
marching across hill and vale
like armies of extra-terrestrial invaders. Robots
with their antennae aloft, their single foot burned black,
burred greybeards frowning through the forests and along river flats.
First I knew the balga on my grandpa's Yenyening farm when amber
chunks of its trunk lay in the woodbox with morning sticks ready to
get the great black stove alight in early mornings. How those
varnished chips of the fallen trunk blazed as our firelighters, filled
the farm kitchen with sweet resinous scent as grandma shifted pots
and pans from the hob, stoked on more whitegum shanks to feed
the crackling fire! Thankful we were, through childhood years, for
this free kindling and forever after that. It was our first gift from
Gondwanaland for us invading immigrants to this Great South Land.
But we were newcomers, for
Noongar people in many
millennia had learned all
and more than we knew.
That, for instance, the soft
growing core at the centre
could be good tucker. Or
that the gold gum from the
balga's trunk helped hold
flint axe heads firm on the
haft; fastened spear barbs
securely on whippy shafts.
But way before this small
birds nested in balga's beard
and beetle and grub found a
place to burrow in the trunk
for sustenance and a private life.

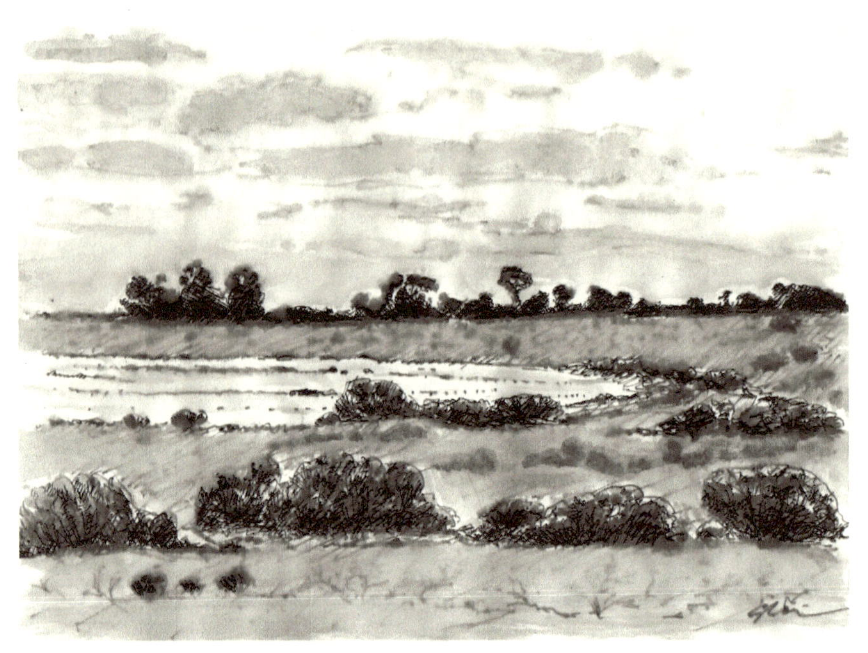

Saltbush and Mulga at Lake Ballard

POWDERBARK WANDOO

Eucalyptus accedens

Stopping on a quiet gravel road
at a small dark green triangle
of reserved bush, I get out
and walk towards this unburnt treasure.

Here I find powderbark wandoos
still standing because the Roads Board
needed gravel pits to build their roads
one hundred years or more ago.

So kept this biodiverse block
with its xanthorrhoea, smoke-bush,
jam trees and termite mounds.
Allowed wodjil and sheoak still to stand.

Forefinger traces your pale bark
and the patina of talcum shows.
Wandoos are a wonder, as Lawrence knew,
saw them as moonlit ghosts in the bush.

His instant midnight lesson there
in Darlington in '22 underlined
him as interloper, the *watjela*
ever to be on the outside looking in.

My grandfather, though sixty years
a farmer near this reserve, came face
to face with the fire of you, my wandoo,
falling asleep at fifty miles-per-hour.

VALLEY IN THE KARRI

Fifty-Second Birthday Poem
Karri (*Eucalyptus diversicolor*)

There's a valley in this forest on a bend
in the road where the karris crowd tall.
It's their pale soaring columns, I recall,
up into the mistiness as if to fend
off lowering clouds, make room for sunlight
to flitter here through dark canopies, gold
shafts in rainforest understorey, to fold
their way down like great wings, fight
and filter via foliage to where you stand,
looking back up to me on roadside here.

Yes, like this karri country's drenched upland
bordering southern oceans, that have raged
unchecked round half the world's rim and rear
on granite shores, our love is fire assuaged.

SMALL BEAUTY AND GRANITE KUNZEA

Granite Kunzea (*Kunzea pulchella*)

> *"High up on the tor itself the grey, gnarled, wizened bushes of Kunzea sprouted pale, knob-like buds, which in the ensuing weeks would open and emblazon the drab, pinched foliage of the bushes with brilliant brushes of cerise flowers"*—Barbara York Main, *Between Wodjil and Tor*

I first saw you sixty years ago
when your red flags were all
frosted with white butterflies
or maybe they were moths.

I'd driven that late spring
an hour or two north-east
of the Avon to Yorkrakine Rock,
exploring that monadnock.

Thirty years later one autumn
I was back and there you were,
still crouched in a crevice
of the granite defiantly.

Your root and branch now
grey, craggy and cramped—like all
survivors on these rock domes,
enduring with old obstinacy.

Is it because you hold close
in your tough green heart
the comfort of your small
beauty's ultimate flourish,

for when Djilba comes,
season of the second rains,
you will send once more
your secret DNA to enact

brilliant defiant scarlet sprays?
Set against velvet green-grey
leaf clusters, they amaze us
as welcome to stone habitat.

Once welcomed, we can belong
to this Yilgarn granite plateau
and stand in the scant shade—
your beauty—small kunzea tree.

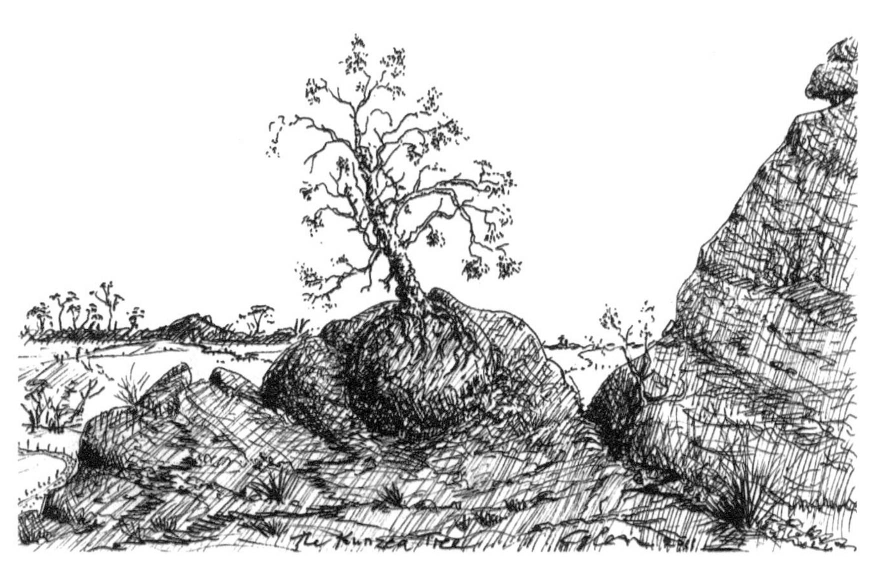

Granite Kunzea in Yilgarn Country

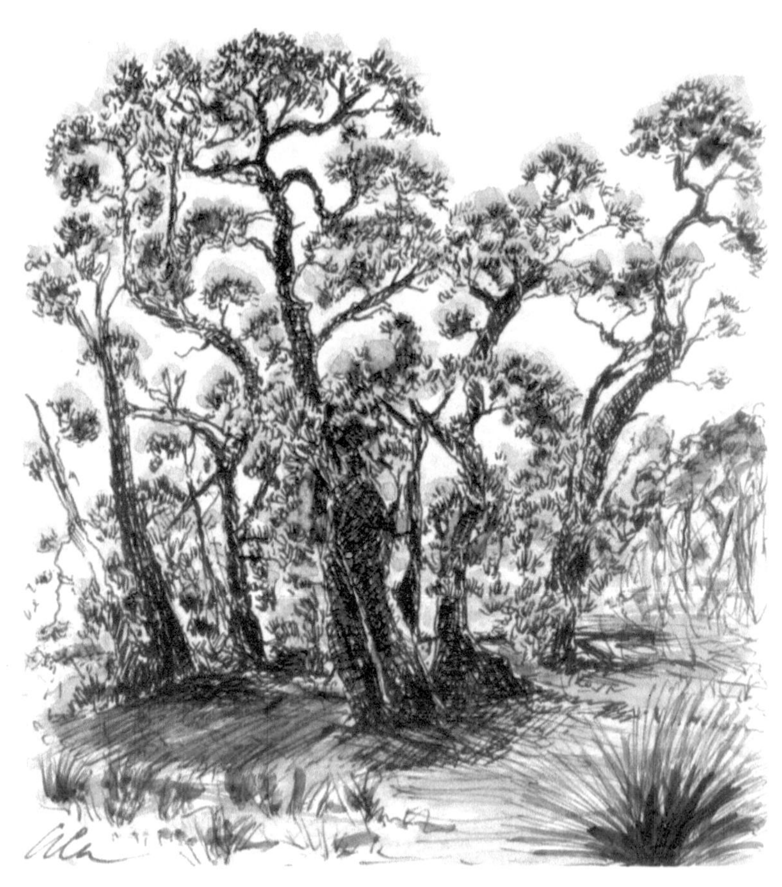

Christmas Trees in Flower

THE FRIENDLY FIRES OF KAMBARANG

Christmas trees or Moodjar (*Nuytsia floribunda)* owe nothing to Pieter Nuyts

Aeons before the Dutch nosed around Noongar
lands, the Christmas tree was known as ghost tree.
Final refuge before departing souls moved on
to Kurannup away over restless western seas.
Moodjar was therefore name for this hemi-
parasite that clamps a white claw round each
host it finds under the earth. And takes sweet
nutrient, be it from tree, shrub or grass.
'Fire tree' the invading settlers called it
in the 1830s or Christmas tree—since
they'd no ken of any of six seasons
of the Noongar year. But Kambarang
was when *Moodjar* bloomed brightest flame, before
the Englishmen sat down to Christmas dinners.

V. Trees Sing

John C. Ryan

RUSTY FICUS

Port Jackson Fig (*Ficus rubiginosa*)

In this province of currawongs and goats, I am watching.
As you cross the fence and enter the field, I am watching.

I am the cornea of this winter field preparing to enclose you.
Tell me, is today the day when the southerly wind is blowing?

Tell me, is today the day when the stacked stones will topple?
I was once water flowing around stone. I hardened in my waiting.

The ribbons of tumbling water calcified to ligaments and bones.
My leaves agreed with the stones, sand, stars and sun watching.

The grazers stave off other trees. Goats manicure this foliate gloss.
When will these inner fruits ripen? My wasps will cease their waiting.

From this rock-strewn rise, I shepherd the slow flexures of seasons.
New families come. Children mature. They leave. I am left waiting.

See my purpling air roots spider darkly as venous blood. Lean in.
Soothe this calloused skin with your touch. Breathe in. Watching.

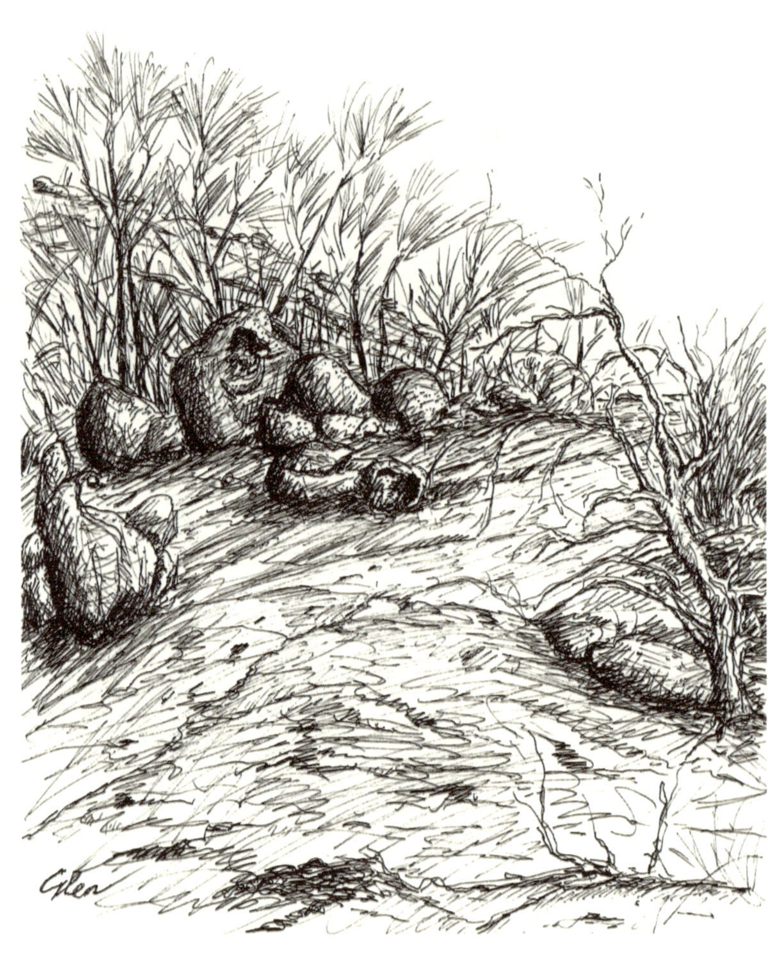

Granite Tors and Sheoaks

CANTICLE

Whistling Pine Tree (*Casuarina equisetifolia*)

sentinel, I dwell in this quadrangle,
gone at dusk as they come, pied currawong
song cleaves the crisp mucous air, I belong
to decibels impelled at odd angle,
accessible to larks who embrangle
along my fuguebrisk updraughted headlong
brawn is borne of golden pollen threadsong,
falsetto at depth of dark tangle.
when by dusk courtyard flush with canticle
and woodswallows croon lunar euphony,
even I blush with moonlight in my cell
and all good hollows of me gush dolce—
again in every sleeping particle,
this harmony awakes and swallows me.

THIS TREE MOURNS THE LOSS OF YOU

Grass Tree (*Xanthorrhoea glauca*)

Once I gave you everything
when once was something between us,
which was ours, yes, but now you think love
is something about everything and nothing,
and nothing about everything but something,
so we live as two obscure anythings in the dust
of beingness because of your thinking you must
live with faith in a thinking anything into being;
I said this once to you—"thought is as old as
things seen from bare ledges we climb
and as vast as chiasmas as these, as
into the everythings of mind
of the things of love, as
loved in time."

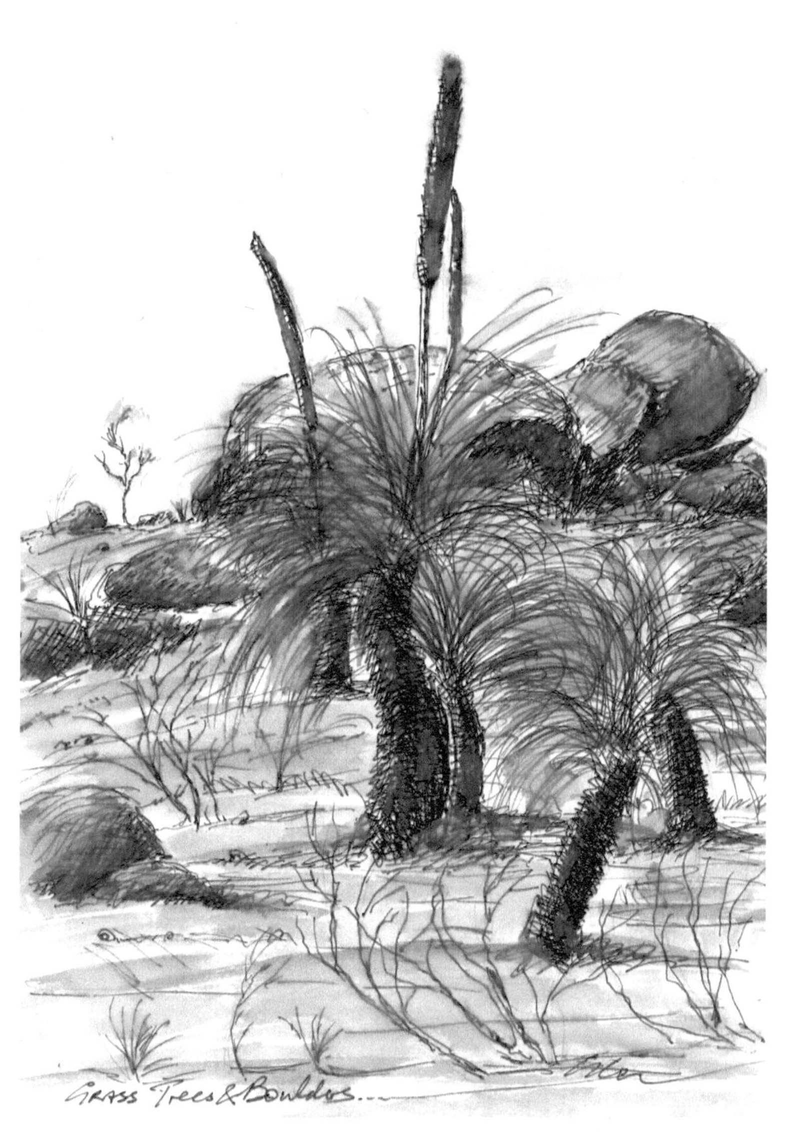

Grass Trees, 'Balgas & Boulders'

YOU FORGOT TOO EASILY

Golden Wattle (*Acacia pycnantha*)

We are not two irreconcilable
beings. Remember that you entered me
then when our bodies seemed infinitely
lured together, magnetic, pliable
in a landscape forbidding, plentiful;
I would never suggest eternity—
your vision could not yield the clarity
needed for me to be believable;
I was not a falsehood then, am not still;
To spur the desire in you to feel
requires of you a gathering will
to receive the world as an unideal.
Who are the darlings you needed to kill?
Where is the fire you needed to steal?

ONTOLOGY OF A PAPERBARK

Swamp Paperbark (*Melaleuca rhaphiophylla*)

i am Drafty's Camp on Warren
torpid all but motionless
i am whoosh of melaleuca
with impulsive woodland zephyr
i am minnows skirting shadows
of steel ramp down to water
i am a tin rowboat bobbing
as family preens the barbie
i am lime-green hue of catspaw
blossoms frilling at their edges
i am the skin of a lizard's back
textured like a golf ball
i am grey-green eyes of boobook
tracking moths in moonglow
i am tawny frogmouth concept
to camouflage in tree crotch
i am forward lurch of joeys
rehearsing their escape routes
i am slinking of the chuditch
patrolling logs and hollows
i am iridescent barksheen
in early evening softlight
i am immersion in the strata
from atmosphere to calcite
*
*
*
*
*
i am an inter leaving.

KING JARRAH OF MANJIMUP

Jarrah (*Eucalyptus marginata*)

A tree is a modular thing—
leaves   break away   bark strips
oft   drift   via torpid billows
of sublimely mentholated air   a-
mass on
treated pine-board apron   on hexagonal
girdle snugly   capillaried 'round root buttressing

King Jarrah respires golden grace with us,
surrendering sinews   capitulating corpuscles
through   five-hundred   torrid revolutions
of igneous drought in un-
forgiving solarium
as cockatoos harmonise   *in medias res*
all of which renders its bole   uncannily perpendicular

for a tree   of *this* kind   in *this* forest.

True, a tree, but also a whirring of consciousness
in self-heal   of blights   and of gouges
initials JE + CG   etched in bark   in hesitant scrawl.

Its arborescence is watchful stillness
that holds on to love you forgot
that you forgot   bear in mind   a tree

always gives back in its taking—

scuffling through hovea & persoonia
a couple pauses to photo she poses
he squinches *clllicksss* by mooncome she'll be risen
in a diaphanous gown of fungus
glowing remember—we are trees in our knowing.

EUPHONIUM: AS THE FOREST WAKES

Red Tingle (*Eucalyptus jacksonii*)

Forests dream in starlight,
        In planet and meteorite.
Skidding off the earth at midnight,
        I dream in limestone stalagmites.

I dream in a coven of trees,
        when quiescent nightflesh is freed
from light's cerebral tendency
        to trust by making trust seen.

Lying foetal on dew-heavy ground
        my fallen brethren, rootbound,
when the subtlest sound
        of frogmouths comes round us.

And sets fire to half-conscious images
        of Kakadu   Xanadu   Kathmandu,
shrieking black cockatoo scrimmages.

In the forest, I wake with the wrens,
        with itching of rashes, resounding clashes,
and nervous clucks of guinea hens.

In fact, a tree, like me, *is* a dream
        lucid   elusive   ever reclusive
at the knotty end of what's known,
                or so it seems.

VI. Tree Tales

Glen Phillips

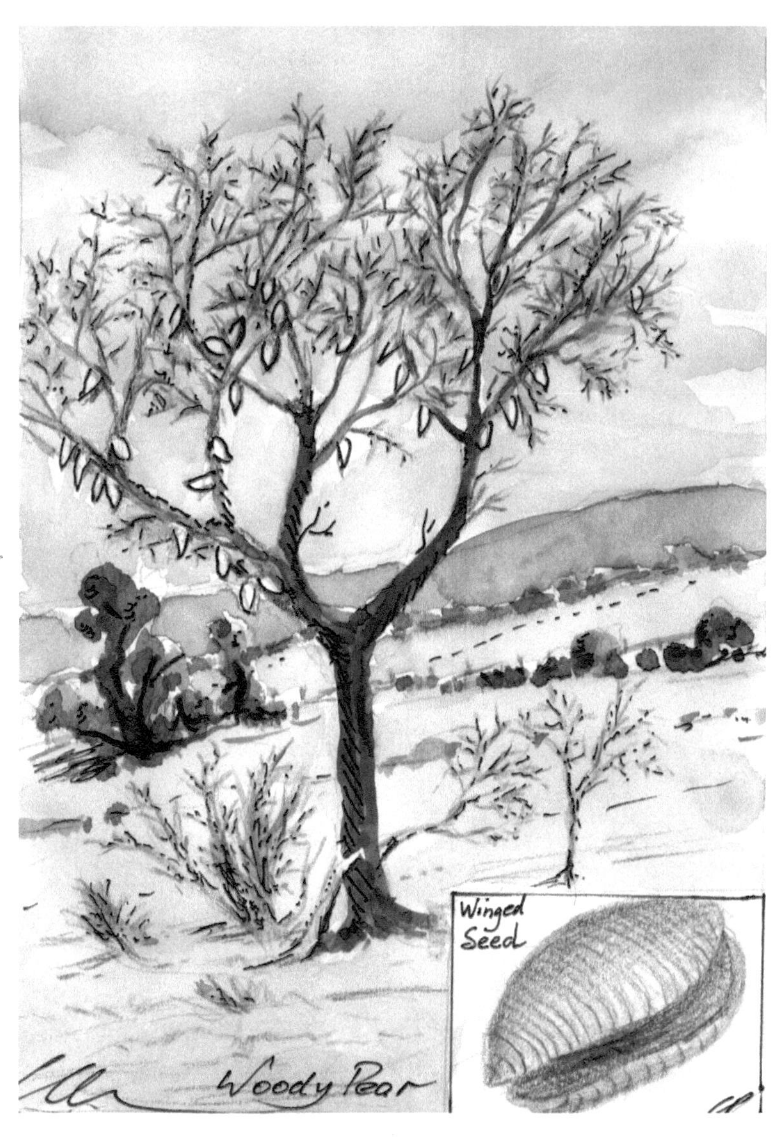

Sandplain Woody Pear

WINGED SEED SONG

Sandplain Woody Pear (Danja, Dumbung, *Xylomelum angustifolium*)

Long ago I was led in the Yilgarn
to the fabled tree of the winged seed.
Danja some called, it and author
Katharine Susannah Prichard mentions
such in *Winged Seed*, that Goldfields
book. And as a symbol, this gauze
aileron, shed from the woody pear,
is spirit of flight, of ideas, re-planting
of cultures and births. Like those
trees with lignotubers, it can thrive
after searing heat or bushfires even,
when the 'pears' crack open wide
and spill their seed to blackened earth.

It was someone, dear to me, though
now deceased, who led me first to
a *xylomelum* tree. Showed me
its strange globes of wood that need fire
to launch their secret ways in flight
to leave parent tree. Barbara York Main
it was, of Yorkrakine, who became *my* winged
seed leading me through these birthland trails,
learning to name and know each place
of wodjil grove, plover, bobtail skink
and trap-door spider. Each flower, shrub,
tree and waft of smoke-bush, all in sight
of granite tor. And with winged seed songs.

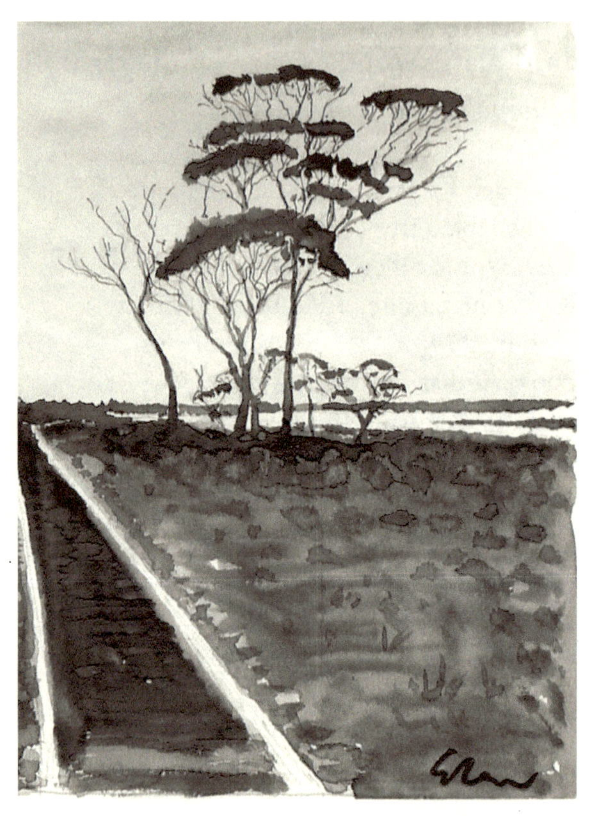

Railway Line and Salmon Gums by Salt Lake

SALMONOPHLOIA SONNET

Salmon Gum (*Eucalyptus salmonophloia*)

"None so elegant and gracile as those trees"—Elliott

Best from a distance—where the eye can dwell
on the soar and arc of each bole and bough—
you cannot deny that they preside now
over lesser heath and hakea and wodjil.
Foliage is pencilled crescents defining well
their slow march beyond lake margins here.
They settle back satisfied their bronzed sheer
presence marks each as a proud sentinel.
Alas, your offspring are few and years passed
before humankind learned to propagate
from your seed—mystery you well preserved
for long, keeping that secret held so fast
in your heart's myriad genetic template.
But at last, discovery we deserved!

TRUE WEDDING GUEST

Barren's Wedding Bush (*Ricinocarpos trichophorus*)

Your white stars deck you finely—
as springtime wedding celebrant. Seen
more and more these clusters, divinely
bowing to suspend a snowy surplice
upon your cassock's verdant green.

Enough for multiple matchmaking
in southern woodlands—Albany west
to Esperance east. And taking
honour to a thousand bedecked brides
to preside over each true wedding guest.

And for each bride too, enough snow
for distant mountain tops seems here;
yet wonder how, out of such flat low
scrublands and rock massifs—Manypeaks
to Cape Barren—manifold shrubs appear?

UNIDENTIFIED FLYING IMAGE OF A POEM

Swamp Paperbark (*Melaleuca ericfolia*)

stumpy swamp.
old forested
silver- in
grey bark
arms paper-
of a

And in the
cragged
limbs
of melaleuca
and thunder of
the cicada hosts
in the deep bracken-
infested aromatic creek
is deep deep-scented tang of rainforest under-brush
wet water-rounded dark stones in
sluggish tannin-stained pools and detonations
of bullfrogs—host to immutable cycle of mutability: conception to decay,
birth to death.

You tease out the oakum of these momentary images
on the workhouse yellowed pine table until fingers ache.
You spread the pale paperbark pages and dip a nib in ink.

RED RIVER GUMS OF GREENOUGH II

Eucalyptus camaldulensis

Leaning in the wind or with the wind?
Steady summer and winter southerlies
at Greenough are fellow travellers with
red river gums mutely bending low
for you and me who stand here also
on alluvial plains. When inclined
we also find direction, bending minds
like these trees to find a way north.

But maybe trees, almost on their knees,
are pointers for us, begging us to keep
our heads down in obeisance to age?
On bended knee are we also mendicants
inclined to reach out to them now? I stroke
a smooth bole or bough in deep deference.

KURRAJONG LOOK OUT

Desert Kurrajong (*Brachychiton gregorii*)

Climbing trees is child's play
they say. In my flat birthland
if you want to glimpse far off
pastures green or even greener
gravity must be defied. For
great heights demand high trees
when there are few high hills
and even fewer mountain tops.

Our western desert kurrajong,
named by explorer Gregory,
was a favourite tree when
I was old enough to climb
the tall one in our front yard.
My green ladder. It glimpsed
for me the far blue range
we called 'Saddleback Hill'.

Beneath after tree flowering
we found the quaint seedpods
shaped as brown boats with
seeds like sailors all in a row.
But the big boys at school
warned that the thistledown
round the seeds would cause
intolerable itchiness if disturbed.

In the greying landscapes
of childhood, looking back
now, I remember kurrajong
tree so bright all summer
against dun grey-green wheatbelt
scrublands of sheoaks, mallees,
dryandras, blackbutts and jams.
Ever upwards since I've climbed.

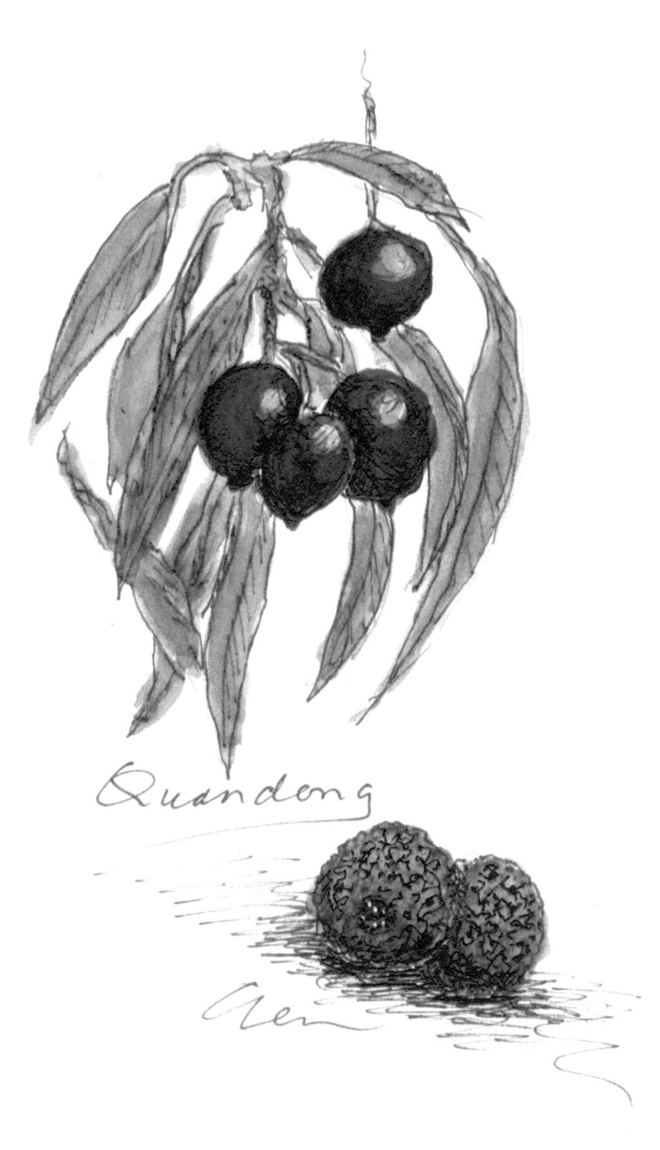

Quandong Fruit

CONTEMPLATING THE QUANDONG

Santalum acuminatum

Driving the Yilgarn we always looked to see
that pale green patch among dun green. Yes,
a smaller parasol amongst casuarina and gum
signalling the native peach or plum. Whatever.
We knew it best as the quandong! We saw first
the sand-coloured nuts, our schoolyard substitute
for scarce glass marble taws. Each wrinkled as
paleo-pygmy's gonads they gave no hint then
of the tart red fruit of this native peach when
tasted in quandong jam or fresh bush plum pie.

But what of this *Santalum acuminatum* out
in its woodlands habitat? They say it belongs
to the 'hemiparasitic' family, plants whose
art is to depend on a host—another tree
or bush, or even a clump of grass to stay
alive. Lucky for the roaming emus, anyway.
They gobble the *wolgol* whole and obligingly
deposit nuts far out from the parent tree
where those rock-hard seeds can soften
and sprout—no competition then, you see.

BATTLE CRY FOR THE OLD GREEN AND GOLD

Acacia spp.

Acacia, acacia,
what'll we do with you?
Your fool's gold foams over
woodlands and farmlands
from western wave-wrack to Pacific rollers.
Oh, the Aussie green and gold we flaunt
on the playing fields of mostly
the Western World and even further on!
That's what we use wattle for.

Acacia, acacia,
in all your devious leaf forms
there remains your green-gold genius.
And then there are your seed-pods.
What'll you do when they explode
in the sun and spray seed over
granite domes and well-grassed swales?

Acacia, acacia,
once we daubed your stems with clay
to make farm-houses. Once we sported
sprigs of flowering wattle
on ANZAC Day* lapels. Green
and gold is battle flag for acacia
in the (what'll we call it?) great internecine
struggle for survival of all plants.

** A day of remembrance celebrated in Australia and New Zealand to honour those soldiers who died at Gallipoli in World War I.*

DRYANDRA DREAMING: MALLET, WANDOO AND RIVER GUM

Eucalyptus camaldulensis et al.

Profuse spring eucalypt blossoms
seem to boil off in flying spume
of pollen to the skies. Join swarms
of insects rising from the earth.
At this place of
ochre pits, powderbark wandoo
and florets of dryandra among
the darker mallets, see darting glimpse
                    of fleet numbats.

My mind flashes
back years to brown bodies sprawled
on the Hotham's sandbanks, under
a late winter sun.
But then grey drift
of clouds darkens the scene and I feel
cold winds huddle the honeyeaters
and wattlers on branch and bough.
Sorrowfully it seems, as first
drops of rain come.
Begin the trudge
back up the hill from the great
river gums and those fuming fields
of everlastings and dampiera
                    to take the dusty road
to Cuballing and Popanyinning
once again.

And at Karping
Bridge see these days those sandbanks
stretched along the salt-ruined river
where never again will bodies
plunge under shade of river gums
into the cool sweet waters
and plunge again.

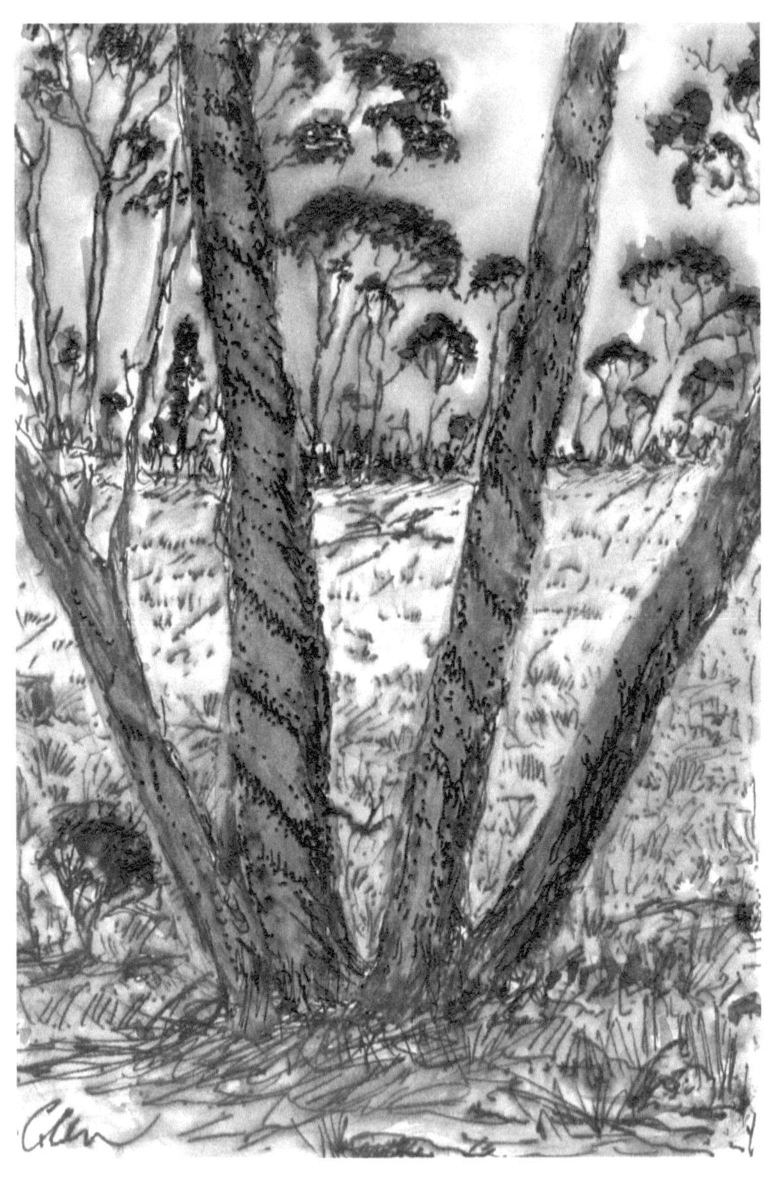

Eucalyptus salubris

GIMLET

Fluted Gum (*Eucalyptus salubris*)

When I first heard your name
uttered by farmers and clearers
chopping out tracts
of our grandfather's Western Woodlands

I could not fathom
origins of your twisted name
for I thought more of penetrating
steel tools drilling

in the grasp of carpenters.
But then was shown among
blackbutts and casuarinas
your manic twisting green trunks.

What tortures twirled these mallees?
Was it suffering salinity
or surviving inland biospheres
here in these arid parts?

Perhaps in youth you gimlets blindly
followed progress of the blazing sun
circling the sky each day from the east,
your craning necks winding your trunks?

Otherwise, like the rest of the trees
you would grow straight and tall
only branching to provide your roots
with a private patch of shade.

Gimlet, we stand in awe to salute
you as a worshipful dervish
of such semi-desert lands,
whirling with this spinning earth!

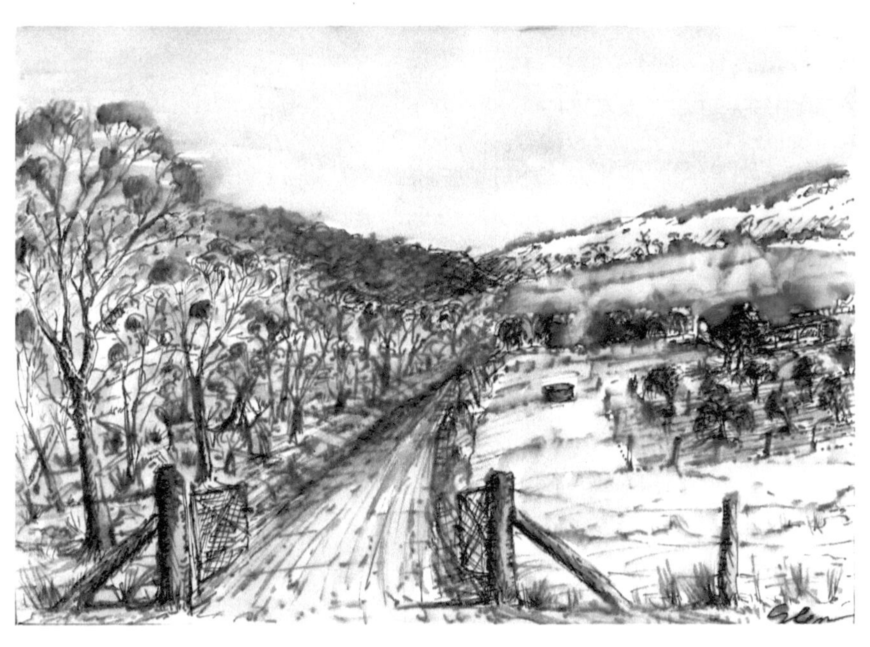

Jam Tree Country Near Beverley

Botanical Index

Botanical Notes

Desert Kurrajong (*Brachychiton gregorii*) was named after Sir Augustus Charles Gregory (1819–1905), an English-born Western Australian explorer.

Firewood Banksia (*Banksia menziesii*) was commonly cut down and used as firewood in the Swan River Colony (later Perth, Western Australia) from 1829 onward.

Gimlet (*Eucalyptus salubris*), also known as Fluted Gum, is a small gum tree prevalent in inland semi-arid areas of Western Australia. Its common name relates to the resemblance of the twisted trunk to a carpenter's gimlet, a tool for drilling holes in wood.

Quandong (*Santalum acuminatum*) wood is known to be ideal for shaping milken or baby baskets for the Ngalia people of the Eastern Goldfields.

Salmon Gum (*Eucalyptus salmonophloia*) is characterised by its brilliant red-brown bark.

Swamp Paperbark (*Melaleuca ericfolia*) is a tree of the myrtle family first described formally in 1797. Its bark is used by Aboriginal Australian people as roofing, as blankets, and for huts, paintings and medicinal purposes.

Western Australian Christmas Tree is named after Dutch Captain Pieter Nuyts who explored the southern coasts of Western Australia in 1627. Part of the area east of Albany was named Nuytsland and, in later years, the vividly flowering Christmas tree or Moodjar was given the botanical title of *Nuytsia floribunda*.

Willow Myrtle (*Agonis flexuosa*), also known as peppermint, is found along the southern coasts of Western Australia.

Wollemi Pine (*Wollemia nobilis*) was discovered in 1994 in a temperate rainforest area of New South Wales. The tree was previously known only through fossil records. A Lazarus taxon is one that disappears from the fossil record, then reappears later.

Artwork

—by Glen Phillips

www.ingramcontent.com/pod-product-compliance
Lightning Source LLC
LaVergne TN
LVHW090955080826
845145LV00003B/1020

9781936671670